International Space Elevator Consortium   ISEC Position Paper 2014-1

Space Elevator Architectures and Roadmaps

International Space Elevator Consortium
Spring 2015

Michael "Fitzer" Fitzgerald
Robert E "Skip" Penny, Jr
Peter A Swan
Cathy W Swan

A Study for Progress in
Space Elevator Development

Space Elevator Architectures and Roadmaps

Copyright © 2015 by:

Michael "Fitzer" Fitzgerald
Robert E. 'Skip' Penny, Jr.
Peter A. Swan
Cathy W. Swan

All rights reserved, including the rights to reproduce this manuscript or portions thereof in any form.

Published by Lulu.com

Pete.swan@isec.org

978-1-329-08717-0

Cover Illustration:
Front – chasedesignstudios.com
Back - TAI

Printed in the United States of America

Preface

The International Space Elevator Consortium vision is to have:

> *"A World with inexpensive, safe, routine, and efficient access to space for the benefit of all mankind."*

One of the principle elements of our action plan towards an operational space elevator is to conduct year-long studies addressing critical topics. This year, ISEC has chosen to challenge the many, diverse, space elevator concepts with a structured approach at architectural development leading to a series of technological roadmaps. Indeed, ISEC understands where the technologies are today and where we would like to be: Initial Operational Capability. However, the process between those time-oriented events must be clarified and illustrated. The goal of this study team is to develop a solid approach to take the space elevator community from today, to a recognized future destination:

Our Destination
The Initial Operational Capability (IOC) contains two space elevators, with separate Marine Nodes, Apex Anchors, 100,000 km usable tethers, and a single Headquarters and Primary Operations Center.

The authors of this report wish to thank:
- the members of ISEC for their support
- contributors to this report for the dedicated efforts
- the attendees of the 2014 International Space Elevator Conference

Signed: *Robert E "Skip" Penny, Jr*
Vice President ISEC

Insight

For the last few months I have been lucky enough to work with some people who have seen the Space Elevator. That might seem odd to say; but, in truth, Pete Swan and Skip Penny have a clear vision of a Space Elevator. In the time cited, I sought to query them to understand how the rest of us could get there, get to what they see, what paths can be taken, and what challenges might we all encounter along the way. This process, called roadmapping, allowed the three of us to understand what needs to be done, and in what order, over the coming few years. This study report should take you along as well.

Michael "Fitzer" Fitzgerald
President, Technology, Architectures and
Integration, LLC.,
Spring 2015

Executive Summary

The study team took on the challenge of explaining a path to develop a major revolution in space transportation, the space elevator. Presently, there are three "validated"[1] architectures addressing space elevators:

- Dr. Edwards' NAIC study, with refinements in his 2003 book.
- International Academy of Astronautics Study Report, 2013.
- Obayashi Corporation concept at Heads of Space Agency Summit, Jan 2014.

Each of these have strengths, weaknesses, proponents, detractors, and most importantly – great potential. This study hopes to refine a process to significantly move the development of this mega-project towards its Initial Operational Capability (IOC). This architectural process, to be refined during the first part of this study report, will not be dependent upon a single past architecture. The process will look at many layouts and approaches to present the future within a disciplined, and proven, methodology.

During the exploration period of this study, it was determined that the Space Elevator Architecture would be composed of five delineated segments. These delineations helped the study team to agree on what was included in the Architecture; and, what was not. The team determined that the Architecture would be a growing entity; so, the destination was the Space Elevator Architecture when its initial operational capability was achieved. That is, we envisioned the Architecture when it would be "open for business." We call this the "IOC Architecture."

The five Segments of the Architecture are:
- Marine Node segment,
- Tether segment,
- Climber segment,
- Apex Anchor segment, and
- Headquarters / Primary Operations Center

These segments are more completely defined in the latter portions of this report; but, suffice to say, all aspects of an Architecture AND all aspects of the total enterprise are included in these five segments. We characterize things this way to

---

[1] In this context, validated architecture means publication of a full Space Elevator architecture (addressing all major elements), discussions at major venues and conferences, defense of major approaches and refinement of concept during the process.

attain the proper level of completeness for the sake of solid systems engineering activities that will be followed as the Architecture is assembled.

The team examined the five segments and identified the most challenging technology or engineering feat that – if and when successfully conquered – would show that segment was indeed viable and would be dependably part of the Architecture. It was determined that each of these five challenges should be treated as if they were prototypes. These prototypes – the challenges - are to be resolved by demonstration. Solid test engineered experiments which show that engineering and performance standards have been, or can be, met. These standards are derived from the performance expectations of each segment. Each segment's final demonstration will be preceded by a carefully selected set of smaller prelude demonstrations . These prelude demonstrations will examine the engineering approach and the validity of the test approach for each segment. Technical risk management at its finest. In this manner, the sequence of tests in the prelude demonstrations represent both a test campaign AND a roadmap for each segment of the Space Elevator Enterprise. Each must be successfully traversed. In addition, we must learn during our journey that the success criteria must be set early and the related standards must be harnessed as requirements in order to reach the Architecture.

We expect that successful completion of the five segments' challenges will reveal that certain technologies need to be identified, matured, and included. In fact, in some cases, technology alternatives must be identified to ensure satisfactory prototype testing or demonstration completion: a selective dual path strategy. As a specific engineering product is foreseen, technology maturation roadmaps will be traversed for each segment. Most will be traversed early. Some will be traversed concurrently with the test events that preludes challenges and prototypes. And some will be traversed later forming the basis for post IOC insertion.

As the study is drawing to a close, the team believes that we have agreed on a series of finite pathways to the future for each segment. The pathways have been identified and are being characterized as getting the Enterprise to the implementation planning phase. This is the phase in which the Elevator Enterprise will be designed, built, tested, and initially operated: a baseline capability of our future. What we have done certainly mandates that we need to build some technology maturation plans and conduct engineering validation efforts to acquire the skills, knowledge and confidence to enter the next phase. This study report lays out a series of roadmaps that will show us the future and provide an insight into "how to" arrive at our destination: the IOC.

    That seems a bit like we are
        ..."ready to begin to start"
        ... but indeed ... we ARE ready to begin to start.

Table of Contents

1. Introduction 1
   1.1 Historical Architectures 1
   1.2 ISEC Study Process: 2
   1.3 Chapter Layout 3

2 Getting to a Space Elevator Architecture 4
   2.1 Prolog 4
   2.2 Circumstance to Strategy to Plan 4
      2.2.1 Circumstance and Strategy...5
      2.2.2 Strategy to Plan..5
      2.2.3 Transformation to a plan..5
   2.3 Getting to the IOC Architecture 6
      2.3.1 The Implementation Plans..11
   2.4 Technology Development Approach 11
      2.4.1 Prolog..11
      2.4.2 Technology Development...12
      2.4.3 From a Technology Development Roadmap to the ISEC Technology Development Plan 13

3 Architecture Approach 16
   3.1 Prolog 16
   3.2 The Anatomy 16
   3.3 Breakout [Structure, Basis for Roadmaps] 18
      3.3.1 The Architecture's Breakdown Structure...19
      3.3.2 The Government's Role 20
      3.3.3 The IOC Architecture - the Destination 20
      3.3.4 Integrated Testing 21
   3.4 Next Five Chapters: Segment Roadmaps 23

4 Marine Node Roadmap 24
   4.1 Introduction 24
   4.2 Marine Node Segment Definition and Mission 26
   4.3 Marine Node Pathway 27
   4.4 Marine Node Culminating Demonstrations 28
      4.4.1 MN #1 Tether Terminus 28
      4.4.2 MN #2 Position Management...29
      4.4.3 MN #3 Attach and Detach..29
      4.4.4 MN #4 Power..29
   4.5 Marine Node Success Criteria 29
      4.5.1 MN #1 Tether Terminus 30
      4.5.2 MN #2 Position Management...30

  4.5.3 MN #3 Attach and Detach 30
  4.5.4 MN #4 Power 30
 4.6 **Marine Node Summary** 30

5 *Tether Segment Roadmap* 31
 5.1 **Introduction** 31
 5.2 **The Definition and Mission of the Tether Segment – The" Ribbon"** 32
  5.2.1 Tether Segment – Node to Node association 33
 5.3 **Tether Segment Pathway** 33
 5.4 **Culminating Demonstrations** 36
 5.5 **System Model** 37
 5.6 **Tether Segment Summary** 38

6 *Tether Climber Segment Roadmap* 39
 6.1 **Introduction** 39
 6.2 **Tether Climber Segment Definition and Mission** 40
 6.3 **Tether Climber Pathway** 41
 6.4 **Tether Climber Culminating Demonstrations** 43
  6.4.1 Additional Demonstrations 43
 6.5 **Success Criteria** 44
 6.6 **Tether Climber Segment Summary** 44

7 *Apex Anchor Segment Roadmap* 45
 7.6 **Introduction** 45
 7.2 **Apex Anchor Segment Definition and Mission** 45
 7.3 **Apex Anchor Segment Pathway** 50
 7.4 **Apex Anchor Segment Culminating** 51
 7.5 **Success Criteria** 51
 7.6 **Apex Anchor Segment Summary** 51

8 *Headquarters and Principle Operations Center Segment Roadmap* 52
 8.1 **HQ/POC Segment Introduction** 52
 8.2 **HQ/POC Segment Definition and Mission** 52
  8.2.1 Enterprise Operations Center 54
  8.2.2 Base Support Station 55
  8.2.3 Transportation Operations Center 55
  8.2.4 Climber Operations Center 55
  8.2.5 Tether Operations Center 57
  8.2.6 Apex Anchor/GEO Node Operations Center 58
  8.2.7 Organization Chart 58
  8.2.8 HQ/POC Pathway 58
  8.2.9 HQ/POC Success Criteria 59

- 8.3 HQ/POC Segment Summary 59
- 9 *System Integration* 60
  - 9.1 Introduction 60
    - 9.1.1 The Humanity of Engineering Interfaces 61
    - 9.1.2 Interface Management 61
  - 9.2 System integration objectives 62
    - 9.2.1 Tether Stabilization 62
    - 9.2.2 Telemetry Receipt and Analysis 63
    - 9.2.3 Payload Exchange 63
    - 9.2.4 Operations Scenarios 64
  - 9.3 Final Systems of Systems Integration Testing with Operations Checkout 65
  - 9.4 Summary 66
- 10 *Summary, Recommendations, and Vision* 67
  - 10.2 Summary 67
  - 10.2 Recommendations 68
  - 10.3 Future 69
- *Appendix A* *Architecture Attributes* 70
- *Appendix B* *Comparison of Current Architectures* 71
- *Appendix C* *International Space Elevator Consortium* 84
- *Appendix D* *TAI's Architecture and Roadmap Approach* 87
- *Appendix E* *Space Elevator Workshop* 88

List of Figures

Figure 2-1　Initial Transformation 6
Figure 2-2　Continuing Transformation 7
Figure 2-3　Continuing Transformation 8
Figure 2-4　Five Pathways of an Executable Test Campaign. 9
Figure 2-5　Becoming a Plan 10
Figure 2-6　Technology Development...12
Figure 3-1　The Climber Segment Roadmap 17
Figure 3-2　Total Architecture...20
Figure 4-1　Marine Node Concept 1 24
Figure 4-2　Marine Node Concept 2 24
Figure 4-3　Marine Node Concept 3 25
Figure 4-4　Marine Node Roadmap...27
Figure 4-5 Marine Node Pathway..27
Figure 5-1　Tether Roadmap..32
Figure 5-2　Tether Segment Pathway 34
Figure 6-1　Tether Climber Schematic 39
Figure 6-2　Emerging from Protective Box...39
Figure 6-3　Climber Segment Roadmap 41
Figure 6-4　The Tether Climber Pathway..42
Figure 8-1　Operations Centers...53
Figure 8-2 Major Functions 54
Figure 8-3　Notional Operations Center..55
Figure 8-4　Robotic Arm Example 56
Figure 8-5　HQ/Primary Operations Center Organization Chart..58
Figure 9-1　Integration Testing-Birthing the Architecture..60
Figure 9-2　Tether Stabilization...62
Figure 9-3　Telemetry Links..63
Figure 9-4　Payload Exchange 64
Figure 9-5　Operations Scenarios 65
Figure 10-1　ISEC's Space Elevator IOC Architecture 67

List of Tables

Table 4-1　Marine Node Culminating Demonstrations...28
Table 5-1　ISEC's Space Elevator IOC Architecture [chasedesignstudios.com] 31
Table 5-2　Tether Culminating Demonstrations 35
Table 6-1　Tether Climber Demonstrations 42
Table 7-1　Apex Anchor Demonstrations 50

1. Introduction

1.1 Historical Architectures

This study will look at the knowledge developed over centuries and assess the architectures related to a space elevator, defined as stretching from the surface of the Earth to well beyond geosynchronous orbit for balance. Comparison of the published architectures for space elevators will be based upon a few basic criteria:

(1) Publishing (with distribution of concept) which created structured information for significant advances in the development of space elevators,
(2) The engineering level of detail was appropriate for the phasing of each report, and,
(3) The presentation showed engineering as feasible, enhancing credibility for the development of space elevators.

This led to a series of five architectures over the last 55 years. The first two were significant leaps in understanding while the last three showed innovative engineering solutions for the 21st Century:

- In 1960, Yuri Artsutanov presented an approach visualizing how it could be achieved – a big leap from Tsiolkovsky's concept from 1895.
- Then, in 1974, Jerome Pearson resolved many issues with engineering calculations of tether strength needed, and approaches for, deployment. This was once again a leap beyond Artsutanov's work and set the stage for the "modern design" for space elevators.
- Dr. Edwards established the current baseline for design of space elevator infrastructures at the turn of the century with his book: "Space Elevators" [2003]. He established that required engineering could be accomplished in a reasonable time with reasonable resources. His baseline is solid; and, it was leveraged for the next two refinements of this transportation infrastructure concept.
- The International Academy of Astronautics study leveraged Dr. Edwards's design and the intervening ten years of excellent development work from around the globe. The 41 authors combined to improve the concept and establish new approaches, expanding the Edwards' baseline.
- The latest version of a space elevator architecture is the view by the Obayashi Corporation. Their set of assumptions at the beginning of the study [basically human safety] established stricter requirements and resulted in longer development with increased payload capacity.

The first and second appendices will address each architecture in some detail as sequentially improving infrastructure design. The differences will be emphasized as the reader moves forward from one architecture to the next. One key to remember is that when the tether material develops, it will enable the space elevator designers to take the best from each of the architectures and combine them leading to a transportation infrastructure for low cost space access.

This study will show a disciplined approach looking at developing a new space elevator architecture with technological roadmaps. This could lead to a new perception of what a space elevator architecture should look like. This study will advance the level of knowledge of space elevators.

1.2   ISEC Study Process:

The International Space Elevator Consortium (ISEC) has developed the process of picking a key topic for in-depth analysis and then conducting a yearlong study to assess the various aspects of the topic. This focus enables the ISEC to prioritize activities and leverage volunteers with expertise in the chosen fields. The single focus on a topic for a particular year enables the community to bring its strengths together and address the topic at the yearly conference, inside the organization's journal, CLIMB, and through the study process with a resulting report. The topics chosen by the Board of Directors of ISEC have been:

   2010 – Space Elevator Survivability, Space Debris Mitigation
   2011 – Carbon Nanotube Developmental Status
   2012 – Space Elevator Concept of Operations
   2013 – Design Considerations for the Tether Climber
   2014 – Space Elevator Architecture and Roadmaps
   2015 – Design Considerations for the Marine Node

Each study goes through a similar process, such as:

August 2013	ISEC selects topic at Board of Directors meeting
	Architecture announced as the topic at the yearly conference
Aug-Dec	Team formed and initial outline of study topics discussed
Jan-Mar 2014	Specific items discussed, analyzed and studied
Mar-Aug	Paper topics submitted to the ISEC International Conference
August	Focus at space elevator conference on topic
	Mini-workshop brainstorming initiated ideas. See Appendix E
Sep-Jan 2015	Study topics drafted as chapters in the report
Jan-Feb	ISEC Review of Final Document
Feb-Mar	Final review and top level peer review
April 2015	Publish Study Report as ISEC STUDY

1.3 Chapter Layout

Later chapters address specific architecture and roadmap approaches:

- Chapter 2 – Getting to a Space Elevator Architecture
  Describes the approach to develop an implementation plan for each of the five segments; Marine Node, Tether Climber, Tether, Apex Anchor, and the Headquarters – Primary Operations Center. In addition, the chapter explains the approach of a "circumstance to strategy to plan."
- Chapter 3 – Architecture Approach
  This chapter lays out a unique approach to explain how ISEC will lead the development to the destination desired. This destination is the IOC and the chapter shows how the approach gets the organization there.
- Chapter 4 – Marine Node Roadmap
  This chapter will address the development of the Marine Node roadmap.
- Chapter 5 – Tether Segment Roadmap
  This chapter will address the development of the Tether Segment roadmap.
- Chapter 6 – Tether Climber Roadmap
  This chapter will address the development of the Tether Climber roadmap.
- Chapter 7 – Apex Anchor Roadmap
  This chapter will address the development of the Apex Anchor roadmap.
- Chapter 8 – HQPOC Roadmap
  This chapter will address the development of the HQPOC roadmap.
- Chapter 9 – Systems Integration
  This chapter will look at the testing/demonstration approach to ensure complete systems of systems integration.
- Chapter 10 – Summary, Recommendation, and Vision
  This chapter summarizes the study results with a look at the future.
- Appendix A – Architecture Attributes
- Appendix B – Comparison of Current Architectures
- Appendix C – International Space Elevator Consortium
- Appendix D – TAI's Architecture and Roadmap Approach
- Appendix E – Space Elevator Workshop

2 Getting to a Space Elevator Architecture

2.1 Prolog

This chapter is a discussion of how we can get to an implementation plan for each of the Five Segments[2] of the IOC Space Elevator:

- Marine Node Segment
- Climber Segment
- Tether Segment
- Apex Anchor Segment
- Headquarters and Primary Operations Center (HQ/POC) Segment

We are going to get there by having implementation plans for each of the five segments of the Space Elevator Enterprise. There may be more than five segments, as we develop the total infrastructure; but, we'll deal with those later. The point is that we currently see five segments and are trying to visualize these five pieces of the enterprise.

2.2 Circumstance to Strategy to Plan

As the roadmapping team stood at the starting line, we had gathered the several ISEC and AIAA publications, the proceedings from a few of the ISEC annual conferences, and even some international publications. This library had examined and discussed Space Elevators from any number of views. One discussed the Concept of Operations, another discussed the Elevator's high level architecture; and yet another registered an assessment of the technological viability of the whole thing. The team even looked at what was going on around the world. These documents provided valuable, diverse points of view which we sought to distill into a strategy, to converge on an understanding of what we needed to get to the Initial Operational Capability. During our roadmapping effort, it seemed evident that within the envisioned Space Elevator Architecture, and its living Space Elevator Enterprise, a number of new entities and technologies are required. Further, new engineering approaches need to be instantiated and new materials need to be applied as the foundation of the Architecture. In street talk, we need new stuff, new ways to make it, and new ways to operate such things. Where are the Wright brothers and Kelly Johnson when we need them? We are not intimidated by all this, just cognizant of the challenge. Perhaps this is the challenge of our time.

---

[2] These Segments are the result of an initial Requirements Analysis and Requirements Allocation activity of the standard System Engineering approach. Interface Management, another standard function, is addressed in Chapter 9.

2.2.1 Circumstance and Strategy

Nominally, a Technology Development Plan is needed. To get to that plan, the International Space Elevator Consortium will base its path to the initial operating capability (IOC) architecture based upon a technology development strategy of "Show Me." In our view, the "Show Me Strategy" begins with a set of well-constructed simulations and experiments. We feel that a successfully executed strategy will convince funding sources (e. g. foundations) that our vision is worthy. **A funded strategy is a plan.**

2.2.2 Strategy to Plan

Given attainment of sufficient maturity, the efforts are then blended into the program's risk management approach, including approaches to "buy down" the risk at a pace consistent with program execution, schedule and cost. The distinction made here between technological maturity risk and the program's risk management approach is a subtle but important one. The assessment of an item's technology maturity gains its access to the program. If it's not mature, then the technology is not part of the program or goes on to a later "on ramp." Once part of the program, the item goes through the program's risk management program where it's engineering, design and mission value progress are closely monitored.

2.2.3 Transformation to a plan

According to our strategy, the Space Enterprise team will get to the implementation phase by following the pathways identified in the roadmapping process. Until this point, the roadmapping process has focused on engineering, technology testing, analysis and such. In order to transform to an executable effort, schedule and funding must be fused with technical objectives. The Five Pathways must attain a planned sequence structure without losing the technical relationships established in the roadmapping process. Our approach to that transformation is straightforward. We retained the structure of the Work Breakdown that we contrived early in the roadmapping process, inserted the identified tests, experiments, analyses, etc., and then rotated the entire structure, overlaying it onto a schedule. It isn't a trick. It is a reformation planning effort that generates the technical information needed to proceed to design and development.

2.3 Getting to the IOC Architecture

The graphics below depict the parallel Roadmaps that result in the transformation into pathways.

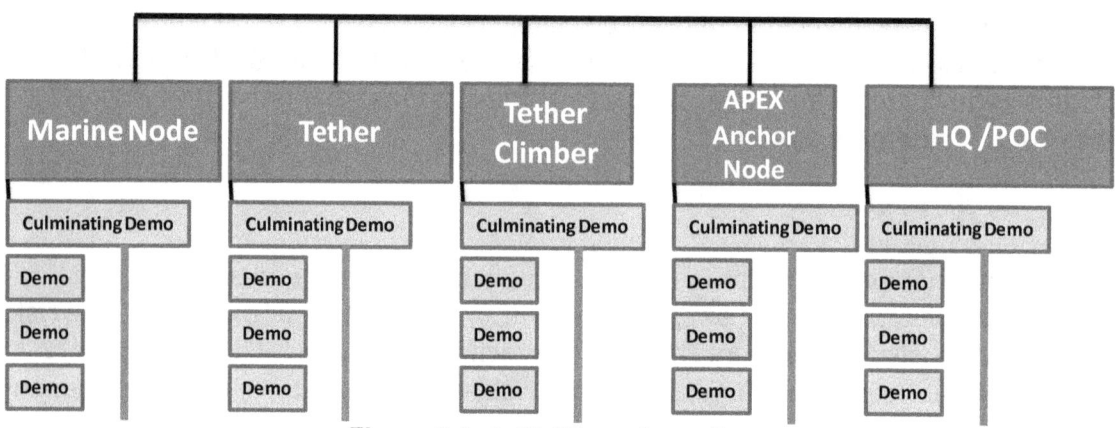

Figure 2-1 Initial Transformation

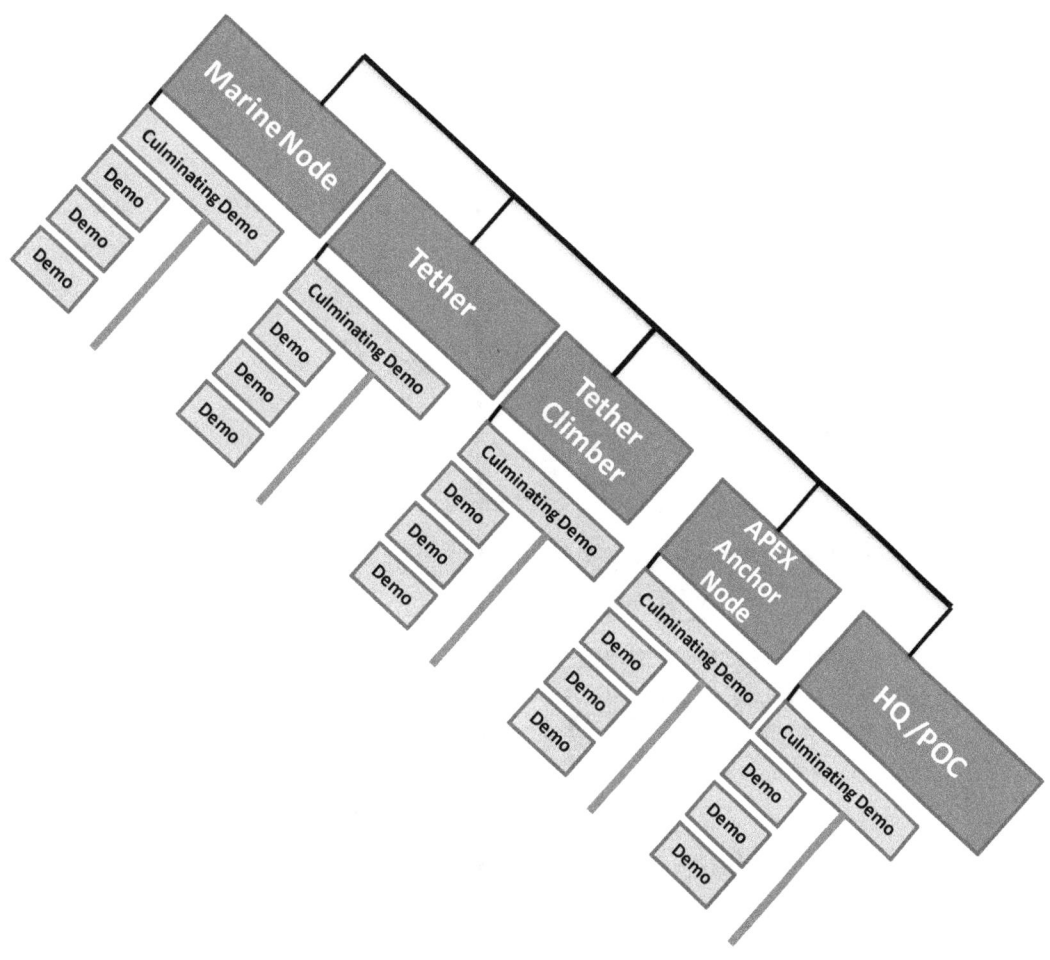

Figure 2-2 Continuing Transformation

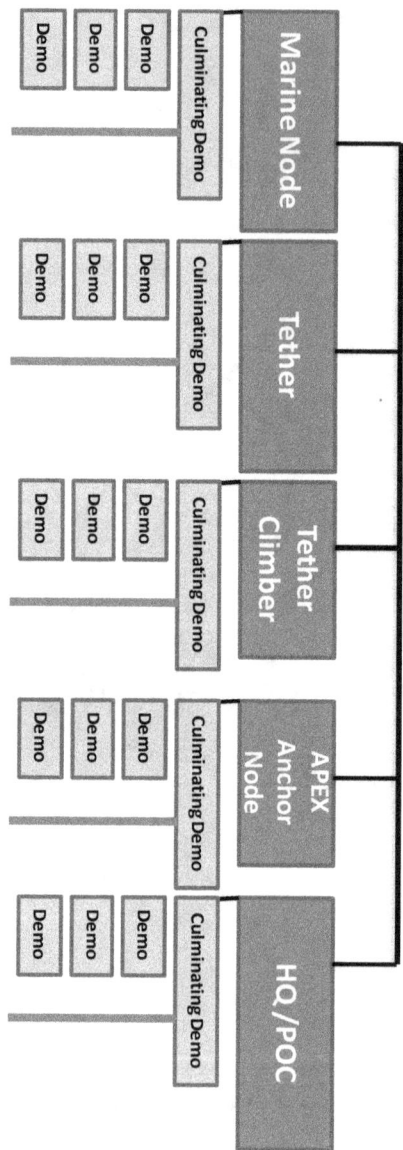

Figure 2-3 Continuing Transformation

The Work Breakdown of the entire technical Architecture of the Space (IOC) Enterprise is reviewed. The duration of various tests and demonstrations are considered in the transformation. This is not drama, it is simply determining how long each of the tests and Demonstrations might take if sufficient resources are available.

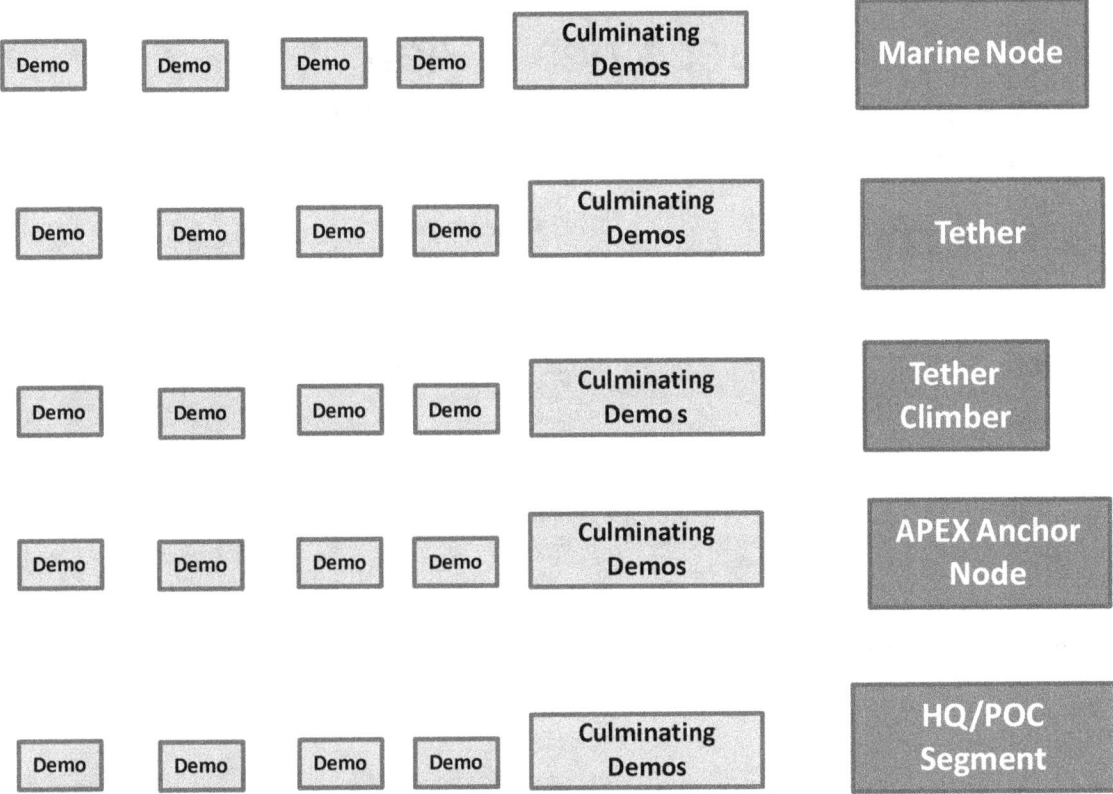

Figure 2-4 Five Pathways of an Executable Test Campaign.

At this point each of the paths must have the strength to move from an idea (the work breakdown structure) to the self-supporting test sequence needed to sensibly arrive at each culminating demonstration The paths must cover the effort to lift immature technologies to an engineering approach for the culminating demonstration and be the basis for the design & development phase. The culminating demonstration should be viewed as a proto-flight, if appropriate. It needs to be much more than a feasibility demonstration, it must be convincing to the investor community, and must be supported by technical content acquired throughout each pathway's execution of multiple test events.

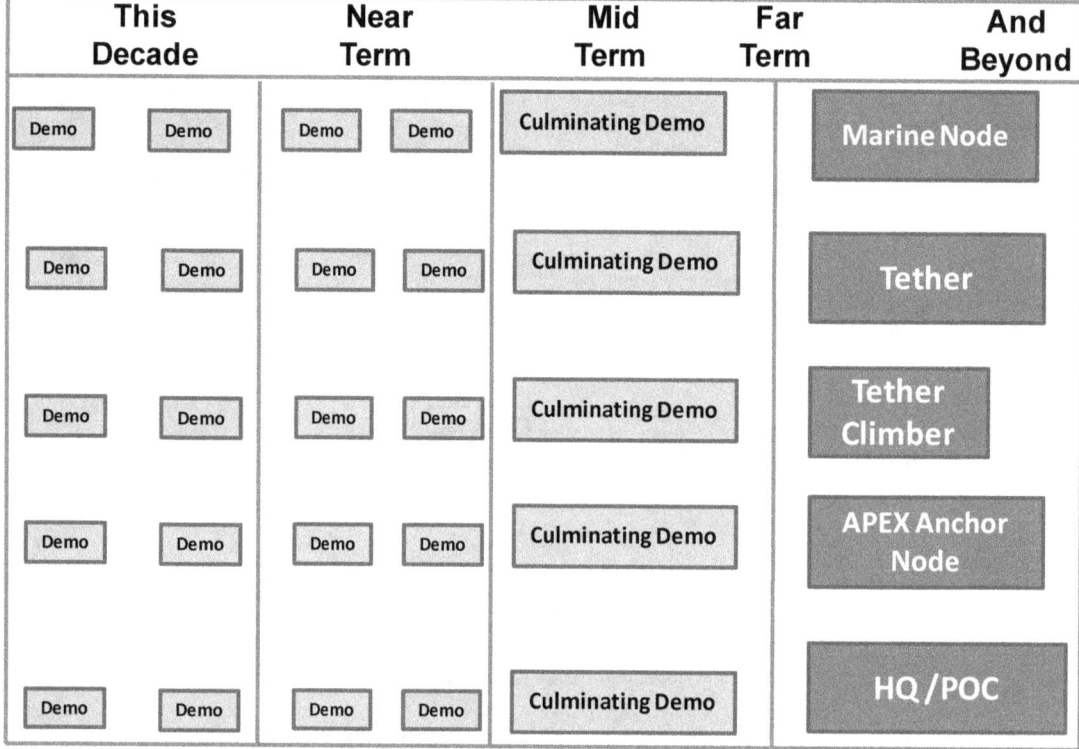

Figure 2-5 Becoming a Plan

As we approach the culmination of the transformation, we will be prepared to execute the campaign of tests, analyses, inspections, simulations and experiments that lead to the culminating demonstration These achievements will posture our team as ready to build the necessary implementation plans: the Five Plans for each of the Five Segments. The initiation of the design and development phase of the Space Elevator Enterprise is here. **Now, we are ready to start!**

2.3.1 The Implementation Plans

Each of the five segments' pathways will be constructed as an executable test campaign. The campaigns are composed of a taxonomic sequence of test and demonstration events that we have been talking about for months. Each event will have entrance and exit criteria and, as we approach the culminating demonstration, each event will have more specific exit standards.

The Architecture & Roadmap (A&R) team saw Architecture implementation planning in three levels of detail, or depths:

- Preliminary Implementation Plan
- Detailed Implementation Plan
- Technical and Mission Review

The implementation plans are substantive items; and, given that they are seen as years in the future, the team portrays that each Segment's Implementation Plan will be modified over the course of time. Build a little, test a little, and build more. In response to the achievements of the preceding test event within the campaigns, and in consideration of the development progress, each step must be compared to the risk state of each segment as the demonstrations progress.

The standards of specification for the three implementation planning documents were not delineated by the R & A team; but, it is clear that the Enterprise Segments must ultimately combine into the integrated Space Elevator Architecture. The Implementation Plans are foreseen as products of the roadmapping process on the one hand and the forecast of technical character on the other. These plans will actually contain libraries of the fundamental System Engineering documents essential for a solid design process to be executed with specification documents, requirements documents, concept of operations, management plans and more. The pathways leading up to an implementation phase gave us the information and the insights to start. Not discussed in this document, but of equivalent importance, are the integration processes that enable the emergence of the single Space Elevator Architecture from the five Segments.

2.4 Technology Development Approach

2.4.1 Prolog

As was discussed in the Circumstance to Strategy to Plan section (2.1), the Enterprise Roadmaps and Architecture team clearly saw the need to mature a range of technologies while also strengthening and validating the engineering approach. This layout is to be used as the Consortium moves toward the design and development of the Space Elevator Enterprise. To service that activity, the R&A team acknowledged a technology development approach to be used within each segment's pathway. The R&A team foresees that each Segment of the Enterprise had its own challenges and would likely need to resolve those challenges in a unique segment manner. The technology and engineering issues facing something at the equator in the middle of the Pacific Ocean are not directly relatable to something at the APEX Anchor at 100,000 kilometers above the equator; and indeed, are in the middle of "Outer Space." As much as the issues are dissimilar, they are the same. Define them, mature the solutions, and determine if we can build something from all of it. The R & A team foresee a specific Technology Development Strategy for each of the five Segments of the IOC Space Elevator. We are going to get there by having implementation approaches for each segment based on the common Technology Development Strategy which follows.

2.4.2 Technology Development

The ISEC Technology Development Strategy follows a tried and true Technology Development Sequence. The ISEC's Technology Development approach will extend the thinking of industry / commercial Technology Plans: seeking to mitigate future risk by buying, developing and /or demonstrating new capabilities.
The generalized roadmap shown in Figure 2-6 depicts a continued inspection of the technical veracity of ISEC progress towards IOC.

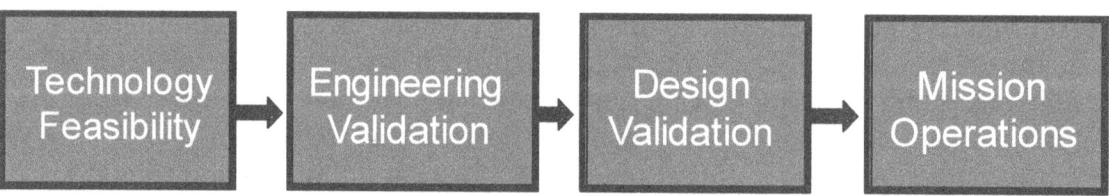

Figure 2-6 Technology Development

The ISEC development plan will be based upon a strategy of a constant and recurring attitude of "show me" manifested in a roadmap of tests, simulations, inspection, analysis, and experiments that reward success with admission to the next set of tests, , simulations and experiments; an iterative approach to risk mitigation. This is the standard evaluation that will be used by each Segment and is portrayed in the Segment's pathways. The nominal Segment roadmap / pathway has two intermediate destinations: recurring Technology Readiness Assessments and Culminating Demos / Prototypes. The final destination is, of course, mission success at Initial Operational Capability (IOC).

2.4.3 From a Technology Development Roadmap to the ISEC Technology Development Plan

The ISEC Technology Development Roadmap becomes the ISEC Technology Development Plan with the identification of specific approaches necessary to execute the Strategy: the four Phases shown.

2.4.3.1 Phase One: Assess Technological Feasibility at the Segment or even lower levels.

This Phase is well underway. The ISEC team has been assessing the technology feasibility situation since 2008. In recent times, the team has begun an open dialog with those members of industry who could be the deliverers of ISEC solutions. Potentially, industry members will be asked to show how the needed technologies are being matured and when they could be dependably available. The ISEC program team will:

a) **Document technology readiness state.** Determine if the technologies are State of Art (SOA) or State of the Industry (SOI) or State of the Market (SOM). "SOA" means that only one industry member holds the critical technology; "SOI" means that a few competent industry members can play; and "SOM" means that the technology is widely available and widely used. This step could start in earnest within the next few years after release of the ISEC Architectural vision and subsequent interaction with industry respondents.

b) **Establish readiness level rationale for all portions of the Program.** Given that the technology availability has been demonstrated (SOA v SOI v SOM ... etc.) the level of readiness can be established for program segment, component or subsystem. Generally, Technology Readiness Levels (TRL) 5 or 6 at the segment level is expected for entry into development. This taxonomy of readiness will be well understood by, and documented in, an official readiness assessment per segment.

c) **Set Success Criteria regarding Engineering Validation.** Prudent acquisition approaches call for an early preliminary design review (PDR). The PDR is an examination to show that the projected engineering approaches are valid. In this consideration "engineering validation" means that we can build it. If the valid technology exists, it can be included in a design based purely upon technology maturity. If a component is SOM, SOI or SOA, or is a TRL level 6, some engineering validation information is needed to get through the PDR process. "Show me" means a lot at this point.

2.4.3.2 Phase Two: Validate Engineering Approaches.

This Phase will begin soon after some publically worthy milestone. The ISEC team will seek a wide range of engineering validation objectives from various members of the industry base. Some efforts might reflect a competitive construct of one segment's envisioned solution, while another segment might be sought in a more collaborative fashion. The Phase two activities are driven by six major activities:

a) **Determine if it can be built:** This is the fundamental question facing the ISEC as it approaches announcements to industry. The ISEC team will describe the segment concepts envisioned and will assess the various engineering approaches being considered by industry. The ISEC team will then ask industry to show how their engineering approach is valid and incorporates the fruits of an ongoing technology maturation effort focused on the Space Elevator.

b) **Examine Industry's technology maturation approaches:** The ISEC team should find ways to review a sample of these roadmaps in industry. It will be clear from the roadmaps that the range and number of needed engineering validation tests are substantive.

c) **Assess schedule & technical risk:** This assessment needs to be very real. Multiple tests and simulations are the path to ISEC program success; and they are the basis of a long sequence of engineering and design judgments. Conducting the numerous tests, resulting in the proper test data and performance insights, is in itself a risky set of ventures. However, proceeding with the program without thorough testing would be beyond risky.

d) **Delineate "On Ramp" Criteria:** Based upon the information on emerging technologies that will not be mature in time, they should be deferred beyond IOC. Setting on ramp targets post-IOC is not simply delay; but rather, a considered approach of when that capability is ("really") needed and whether subsequent maturity and testing will be fruitful.

e) **Set criteria and standards regarding Design Validation:** By the end of Phase Two ISEC should be able to determine whether or not it can build the Space Elevator by determining the efficacy of specific design approaches. Those design criteria and design standards need thorough evaluation for the sake of technology, schedule and/or cost risk. These criteria and standards are to be assessed in Phase Three using design validation information.

f) **Baseline Technical Performance:** By the end of Phase Two, the performance of the envisioned concept can be predicted and will be "baselined."

Phases 3 and 4 are part of the ISEC Technology Development Plan; however, they become the System Engineering Plan for the Space Elevator development program and the path toward mission operations. The outlined activities of each Phase are included here for the sake of completeness. The efforts taken by the ISEC team to get the needed technologies matured (Phase One) and then assessed to be "engineering valid" (Phase Two) must not be omitted as some bureaucratic process. The judgments and efforts of Phases One & Two move forward into the program's subsequent Phases and are amplified by a System Engineering Management Plan, a Test and Evaluation Master Plan, a Risk Management Plan, and other discrete system engineering process efforts. They will ultimately deliver on the promise and vision of those predecessor efforts.

2.4.3.3 Phase Three: Validate Design Approaches

Phase 3 does the following:

a) Service the risk buy down
b) Measure design versus performance baseline
c) Baseline technical performance measures
d) Establish the basis for mission assurance assessments

2.4.3.4 Phase Four: Assess Mission Operations Success

a) Establish performance envelopes for the IOC system
b) Transition risk management program into risk monitoring
c) Examine "On-Ramp items"
d) Baseline operational performance measures

3 Architecture Approach

3.1 Prolog

The A&R team used a unique approach to explain how ISEC intends to lead us all from where we are to where we need to be. That scheme is to capture a roadmap showing where we are and a destination. Over the last year or so, we discussed the technical and operational issues facing the five Segments of the Space Elevator Enterprise envisioned. The team created five roadmaps or pathways taking us from here to there (IOC). The roadmap graphic is really only a work product and is a little messy at this point in time. The work product will refine over the next few years; but, for now, even in its messy nature, we need to be able to read it. Using the Tether Climber Segment as an example, let's take a look at how you can leverage this roadmap graphic. The graph, Figure 3-1 below, is on a whole page as a way to place it in perspective.

3.2 The Anatomy

1 The first step in the roadmap is essentially a statement of where one is. In the mind of the A & R team, this upper left corner is reserved for a statement of where we are. In the coming paragraphs, a description of the Segment and its mission or purpose will describe exactly "Where we are!"

2 This Second step is why we are here. It is a citation of our destination: "An Initial Operational Capability." By having a destination, the roadmap makes more sense as one transitions towards it. Two things happen. The first is that as you get closer, you have a better feel for the steps and processes needed to complete the venture. The second is that you do not become distracted by other, less important, end states.

3 The Third step in each of the five roadmaps / pathways is a documentation of the essential purpose of the functions of the segment. In the case of the Climber Segment, the A&R team saw three key functions of the climber: a) Repairing the tether, b) Releasing satellites at various altitudes from LEO through GEO, and c) Operating above GEO. Each Segment has very different statements at Step #3.

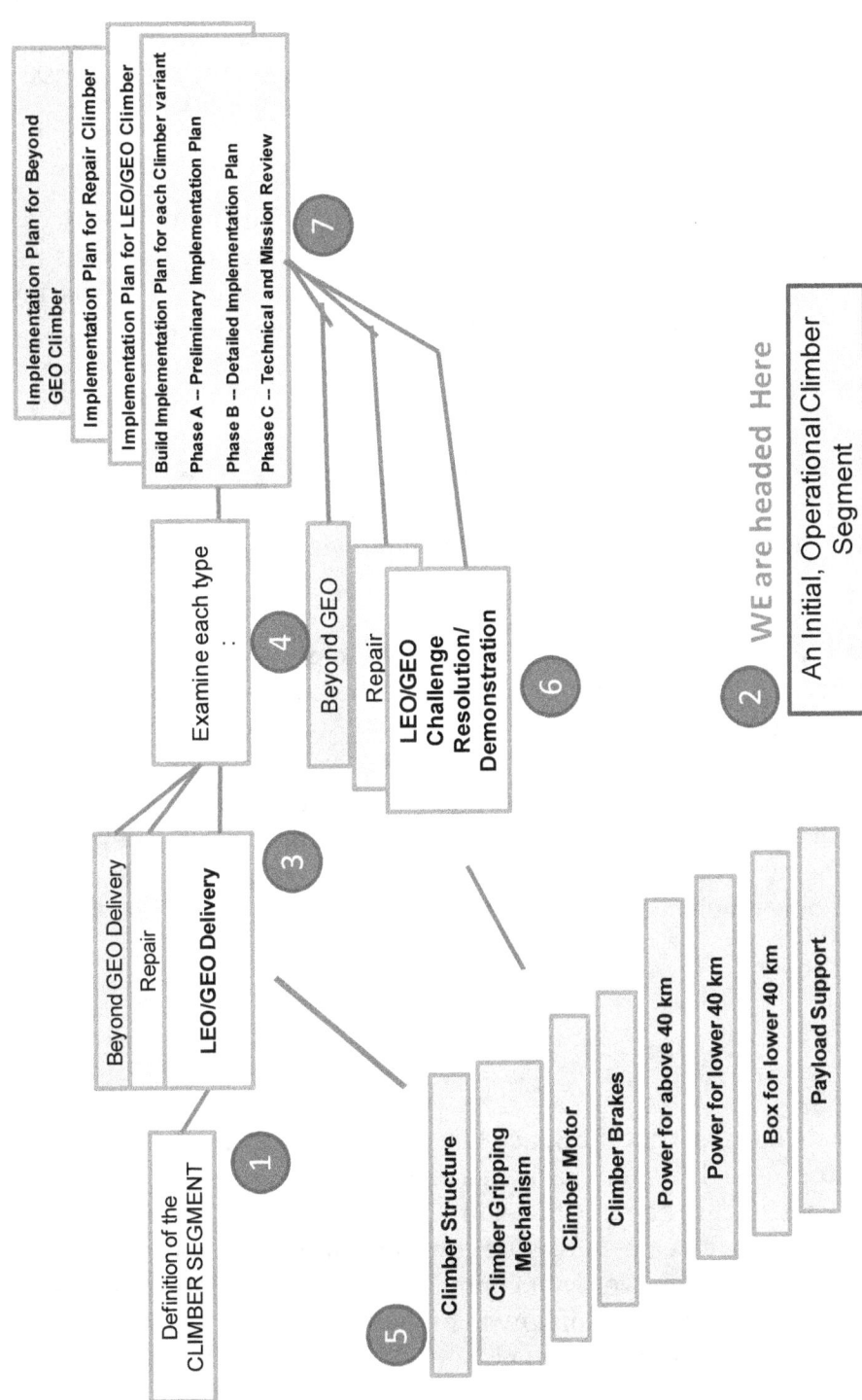

Figure 3-1 The Climber Segment Roadmap

4 Step Four is a valued position along the pathway. It equates to Yogi Berra's admonition that "when you reach the fork in the road --- take it!!" Based upon the topics cited in #2, the team strongly suggests that the Technology Development Plan be used as a reference to determine technology and engineering readiness.

5 Step Five is the beginning and the completion of the readiness assessment process. This distillation of technological and engineering issues reveals the criteria for which progress toward an implementation plan is measured (versus any real and suspected technology and engineering risk judgments levied).

6 Step Six is the citation of the technical content needed within each of the Segment's culminating demo's and the delineation of how (and how well) the Segment is ready to begin execution of the implementation planning process.

7 Step Seven is the start of the building of the programmatic planning package to underwrite and plan the entirety of effort needed to design, build, and test the IOC Space Elevator.

Each of these seven steps is intrinsically in each of the five roadmaps built by the A & R team. All five of the roadmaps are reviewed and discussed in detail in separate chapters. The reader should appreciate two things: how much we learned by the process and how much more we have to learn.[3]

3.3 Breakout [Structure, Basis for Roadmaps]

This section delineates what the Space Elevator is as a whole; and, by implication, what is not the Space Elevator. Early in a development process it is good to determine what you are going to develop. The secret to success is to dismiss what you are not going to develop. After due consideration, the team came to some finite conclusions. The decisions were not complicated. We considered and then used four rules in the determination of how to breakdown the Architecture and how to transform it into an implementation plan:

---

[3] Author's note: Some may question why the fixation on being so thorough even before the implementation planning process. Two reasons: Execution of steps 3 through 6 will be very expensive, but not nearly as expensive as executing the plans themselves, and the investors would expect nothing less.

1. <u>KISS – Keep it Simple, Stupid:</u> The planning and execution of our work over the coming decades is going to be difficult enough Don't exacerbate the effort with a complicated work breakdown.
2. <u>Be aware of the "bite chew ratio":</u> The small Architecture team knows that trying to do too much at one time –"biting off more than one can chew" – has dangerous consequences. Five pieces seemed about right.
3. <u>Avoid stupid things:</u> The best way to get the stupid out of a plan is to go through a process built on risk reduction and risk management. The roadmapping process delivered this with an iterative discovery and risk reduction roadmap made up of a foundational set of solid tests and within each segment.
4. <u>Do NOT depend upon the government</u>: Paragraph 3.3.2 below validates this part of our approach.

3.3.1 The Architecture's Breakdown Structure

Generally speaking the idea of "breaking down" the total system is not very complicated nor very difficult. The real decision is to recognize what you can control and what you cannot. If possible, get the things you cannot control out of your system and organize the remaining into pieces that have functional and location similarity. The result of that thinking is shown below. In simple terms, the A & R team saw that the Total Space Elevator Architecture had five pieces.

- The Marine Node – The place where the elevator attaches to the Earth. It is called the Marine Node because it will be located along the equator in the eastern Pacific Ocean.
- The Tether – The elevator's cable. The tether rises from the Marine Node to the Apex Anchor Node; about 100,000 kilometers above the Earth. The Tether is composed of very strong material; currently viewed to be carbon nanotubes.
- The Tether Climber – The Climber is essentially the elevator's car which carries the satellites to their proper altitude. It climbs to altitude by gripping the tether.
- Apex Anchor – the smart counterweight at the end of the tether. This Node must be able to support thrusting activity to stabilize the tether and have a separate communications node inside the total architecture.
- Headquarters/ Primary Operations Control (HQ/POC) – Every operational systems needs an HQ and an Ops Center

This breakdown is important to portray the pathway to an operational Architecture. As the reader will see in the coming sections, each of these segments will be shown as a destination. If the Space Elevator enterprise is to become real, each of the 5 segments must operate effectively and reliably. Designing and building each segment requires us to overcome a wide ranging set of technology barriers and manufacturing challenges.

3.3.2 The Government's Role

I have read all of the several Jack Reacher novels. I kind of like the guy. He has this healthy skepticism of the government and dislikes those dependent upon it. In one of the novels' plots, the bad guys chasing Jack shot some holes in the wall at a motel where Jack was staying. After Jack chased the bad guys away, the motel owner wanted to call the government to come and fix the holes in his wall. Jack had just saved the guy's life, and then, the guy was looking for government help. WE don't have holes in our Space Elevator walls, and even if we did, I would not call the government in to fix them.

There are many ways in which the governments of various countries could be invited into this enterprise, including fixing little holes. But the truth is, we considered government involvement the other way around. Looking for government involvement only when it offers a real pathway within our roadmap. To this point, we have found no needed capability within the government that must be incorporated in the Space Elevator Architecture. In another part of this report we'll discuss our meeting with the retired Chief Engineer of the Port of Los Angeles. In addition to a host of other questions (which we'll cover later), our new best friend's comments about government involvement were remarkably profound. It was something like "don't expect to get any help and with regard to their rules -- just comply." Sounds like a plan. Hence 'Government Furnished Equipment' is NOT a Segment.

The resulting Space Elevator Architecture Breakdown is shown in Figure 3-2 below.

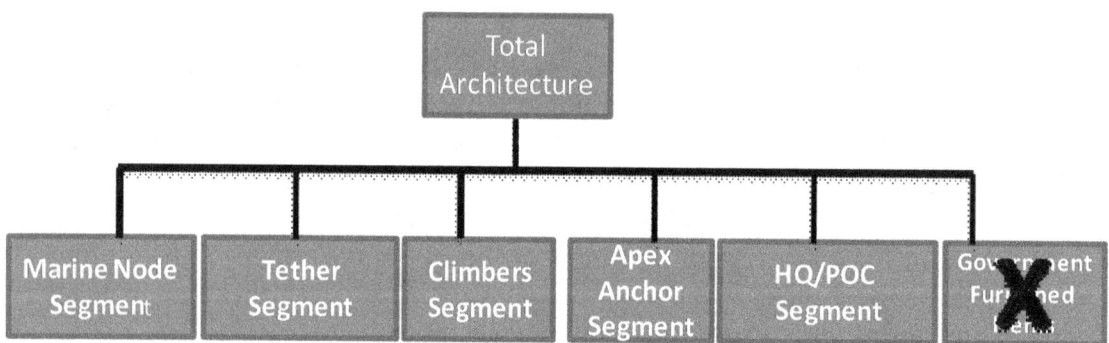

Figure 3-2 Total Architecture

3.3.3 The IOC Architecture - the Destination

The IOC of the Space Elevator infrastructure will consist of the necessary elements for a commercial operator to conduct business. The current concept is something like:
- A replicator space elevator will have a principle purpose of producing more space elevators. This might be significantly less capable than an IOC space elevator as its only purpose is to produce more space elevators [raise tether and apex anchor to GEO to initiate the process again].

- A first commercial space elevator that has a minimal capability to raise payloads to GEO and beyond. This would consist of the HQ/POC Segment which will be the first operational segment as it is needed to implement and control the other segments during IOC activation process. Exactly how the activation at IOC will be executed will be part of the HQ/POC implementation plan. This report focuses on the steps needed to get to the start of the implementation planning activity. That activity is basically System Engineering planning.
- Tether Segment at IOC will emerge from a sequence of tethering activity and will likely have tethers left in place after the engineering validation phase
- Tether Climber Segment at IOC will play a critical and active role
- Marine Node Segment at IOC
- Apex Anchor Segment will probably be relatively simplistic at IOC.

One key is that the strengths of the IOC tether could be less than the optimal commercial – full-up – operational tether. The current estimate for the full-up space elevator is a 31 metric ton capable tether for a climber of 20 metric tons [6 climber, 14 payload]. Obviously, the developmental tether will be minimally capable, while the replicator tether can be built up to handle its set of needs. The operational tether would be much stronger over time. This IOC Space Elevator should be sufficient to handle initial commercial needs (GEO delivery), and expandable to more robust space elevator capability.

3.3.4 Integrated Testing

The word "demonstration" in this study takes on the largest meaning. As such, to satisfy a demonstration that is leading to the "readiness" of a segment, the process would be one of the verification methods refined by NASA.[4] NASA's verification methods are the requirement satisfaction methods being leveraged in this study's discussions. Each of the demonstrations shown in the segment pathways can be satisfied by:

(1) Test-Verification: *Test* is the actual operation of equipment during ambient conditions and subjected to specified environments to evaluate performance.

    (1a) Functional Test: *Functional testing* is an individual test or series of electrical or mechanical performance tests conducted on flight or flight-configured hardware and/or software at conditions equal to or less than design specifications. Its purpose is to establish that the system performs satisfactorily in accordance with design and performance specifications. Functional testing generally is performed at ambient conditions. Functional testing is performed before and after each environmental test or major move in order to verify system performance prior to the next test/operation.

---

[4] The verification methods used by NASA shown in NASA MSFC-HDBK-2221, FEBRUARY 1994.

(1b) Environmental Test: *Environmental testing* is an individual test or series of tests conducted on flight or flight configured hardware and/or software to assure the hardware will perform satisfactorily in its flight environment. Environmental tests include vibration, acoustic and thermal vacuum. Environmental testing may or may not be combined with functional testing depending on the objectives of the test.

(2) Analysis: Verification by *analysis* is a process used in lieu of or in addition to testing to verify compliance to specification requirements. The selected techniques may include systems engineering analysis, statistics and qualitative analysis, computer and hardware simulations, and computer modeling. Analysis may be used when it can be determined that:
    A. Rigorous and accurate analysis is possible.
    B. Test is not feasible or cost-effective.
    C. Similarity is not applicable.
    D. Verification by inspection is not adequate.

(3) Demonstration: Verification by *demonstration* is the use of actual demonstration techniques in conjunction with requirements such as serviceability, accessibility, transportability and human engineering features.

(4) Similarity: Verification by *similarity* is the process of assessing by review of prior acceptance data or hardware configuration and applications that the article is similar or identical in design and manufacturing process to another article that has previously been qualified to equivalent or more stringent specifications.

(5) Inspection: Verification by *inspection* is the physical evaluation of equipment and/or documentation to verify design features. Inspection is used to verify construction features, workmanship, dimension and physical condition, such as cleanliness, surface, finish, and locking hardware.

(6) Simulation: Verification by *simulation* is the process of verifying design features and performance using hardware or software other than flight items.

(7) Validation of Records: Verification by *validation of records* is the process of using manufacturing records at end-item acceptance to verify construction features and processes for flight hardware.

(8) Review of Design Documentation: Verification by *review of design documentation* is the process of verifying the design through a review of the design documentation during the Preliminary and Critical Design Reviews.

3.4 Next Five Chapters: Segment Roadmaps

Space Elevator Infrastructure Segment Roadmaps are shown in the next five chapters to ensure a complete understanding of the individual challenges within each major segment. These are specifically developed to expand on our knowledge base for all segments of a space elevator for the reader. Each segment will be broken down within its own chapter, as follows:

- Segment Introduction
- Segment Definition and Mission
- Segment Pathway
- Culminating Demonstrations
- Segment Success Criteria

As one goes through these chapters, one sees the parallels and the uniqueness of each segment. Even though the pathways are parallel, there is much overlap and interconnection during the testing and demonstrations.

4 Marine Node Roadmap

4.1 Introduction

Some of the ideas for the Marine Node are portrayed with the following three images. The first is from Frank Chase, the second from the Obeyashi Report, and the third is our own image.

Figure 4-1   Marine Node Concept 1

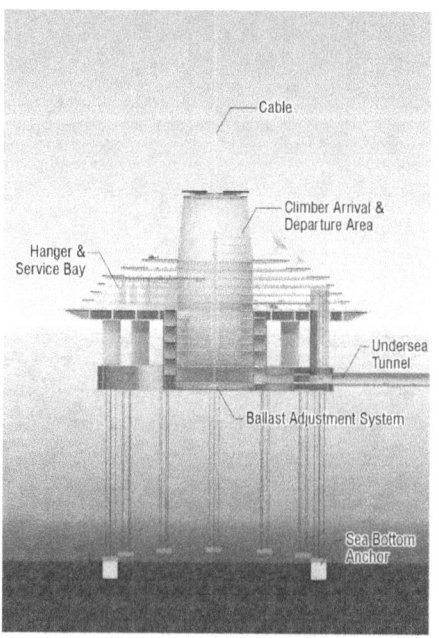

Figure 4-2   Marine Node Concept 2

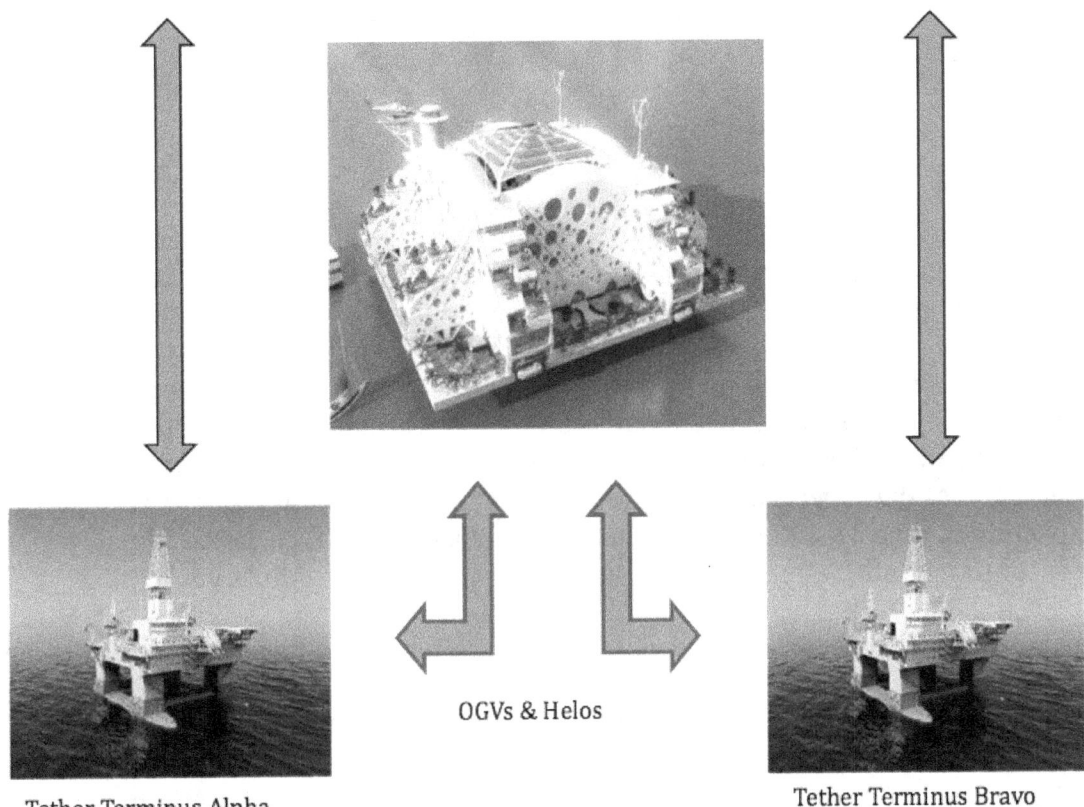

Figure 4-3   Marine Node Concept 3

The Marine Node Segment's path is seen as resolving four primary functions:

- Serves as Floating Operations Platform(s) for the Space Elevator Tether terminus including reel in/out (tension, wind, current, debris avoidance), and position management.
- Serves as a port for receiving and sending Ocean Going Vessels. The OGV's that come and go from the Marine Node are moving Climbers, payloads, supplies and personnel.
- Hosts attach and detach of tether climbers
- Provides care and feeding of crew members including power, desalinization, and waste management

As indicated by the roadmap in Figure 4-4 below, and for the sake of the above four Mission primary roles, an ordered taxonomy of sequenced tests will be conducted. In support of this test activity a wide ranging surveillance must be conducted

25

related to Marine Node site location. Ocean currents, surface wind speeds, water temperature and salinity variations, sea floor geology and weather information will be collected. Ultimately, the validity of the Marine Node will be shown in 4 culminating demonstrations.

4.2 Marine Node Segment Definition and Mission

Basically, the Marine Node provides a location for the tether terminus that can enable safe and routine operations. Its primary purpose is the mating and de-mating of satellites and climbers. This would include stabilizing the tether, moving the tether, loading and unloading of cargo, and local operations support. It is where the climber is prepared and then "sent on its way" safely. This fundamental support of the two-way transportation of goods connects the Space Elevator to the Rest of the World and the Rest of the World to the Space Elevator. The Marine Node will tie together all of the aspects of the terrestrial component to include safety, security, inspection of cargo, loading of cargo to climber, loading climber on the space elevator tether, off-loading climbers, and support to teams in the area.

The Marine Node is a city on multiple floating platforms in the eastern Pacific Ocean. The main element of the Marine Node, the Floating Operations Platform (FOP), will be the size of an aircraft carrier or larger. Secondary elements are now envisioned outlying from the FOP to give the tether terminus a strong base leg for tether anchoring stability. The Marine Node will have living quarters, kitchens and laundries, as well as recreational and medical facilities. It facilitates helicopter landings, local support water craft, and the loading/unloading from larger Ocean Going Vehicles.

The FOP hosts a local Operations Center for management of tether, tether terminus, and platform operations. In addition, the Center supports climber operations including the operations and maintenance of the tether. The FOP vision is just now growing into something specific. The FOP could be tethered or free floating. It could be a deep sea drillship: either a ship designed and built to be a drilling vessel or an older fitted with drilling equipment or, in our case, it could be refitted to perform the functions of the FOP described above. The Marine Node is currently being more completely defined in the 2015 Space Elevator definition effort led by a team with extensive shipping, open-ocean, and port operations experience. The Marine Node Roadmap from that report is in Figure 4-4 below.

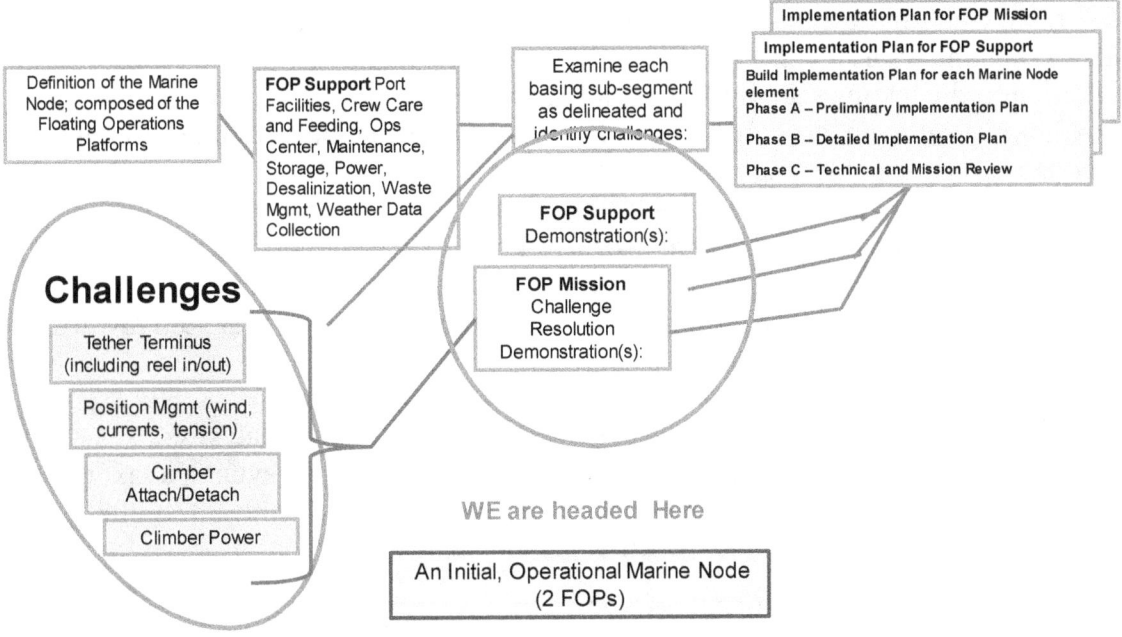

Figure 4-4 Marine Node Roadmap

4.3   Marine Node Pathway

As can be seen in Figure 4-5 below, the effort over the coming time is to traverse the Marine Node's Roadmap based Pathway, getting to the Planning Phase with proven technology and validated engineering approaches. These are achieved with a sequenced series of tests, experiments and simulations (called "Demonstrations" for simplicity).

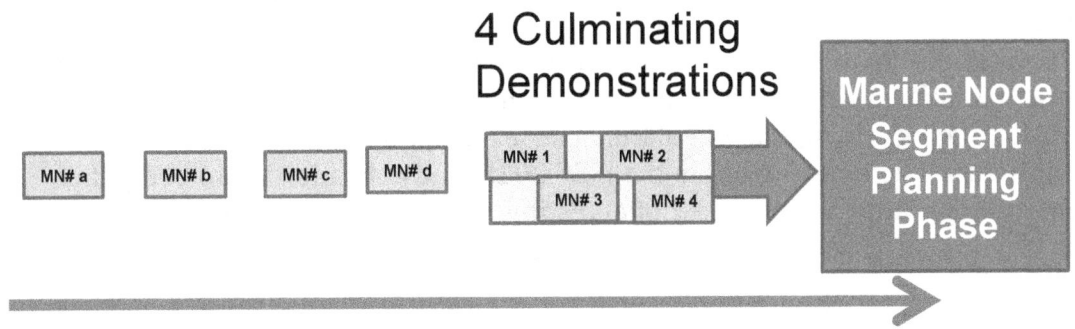

Figure 4-5 Marine Node Pathway

The demonstration line portrayed in the pathway is essentially a test campaign conducted at each step with entry and exit conditions cited as part of the campaign execution. Note that the test campaign will focus on the Demonstrations associated with the challenges shown in the roadmap in Figure 4-4. To better understand the scope of this test campaign, and the specific test events that might be needed, the A

& R team went to the 2014 ISEC Space Elevator conference and asked for feedback about the kind of test events they thought would demonstrate essential progress. In addition to the testing seen by the A & R team, and beyond the testing implied by the Roadmap & Pathway, the conference audience provided a good bit of wisdom in its feedback. The idea of letting months of hard work on one's little project be open to the public for inspection is always a little embarrassing. As it turned out, it was not embarrassing. The attendees were warm and welcoming in their feedback. It was a wonderful, humbling moment and the entire team was proud to have been a part of it. The engineering of the ISEC Space Elevator is underway! The first step toward our destination had been taken.

"Demonstrations" should include a range of tests, inspections, analyses, simulations, and more. Some are likely to be a sequence of test events; a taxonomy of tests. The Term for each is expressed as: Near (2014-2020), Mid (2020-2028), Far (2028-2035). This will apply to the other segments as well.

Term	Demonstration	Dependency
MN-1 Far	Tether Terminus (includes reel in/out)	Needs Tether
MN-2 Mid	Position Management	Needs Tension spec
MN-3 Far	Climber Attach/Detach	Needs Tether
MN-4 far	Climber Power	Needs Climber

Table 4-1 Marine Node Culminating Demonstrations

The Marine Node will also be the base for any number of operational tests during the transition toward IOC. In that sense, the Marine Node holds a unique position as the center of activity for test and demonstration practice and training. Given that the A & R team has established the idea of a minimum IOC system which then grows into something robust, the role of the marine node is still to be defined. That definition will emerge further in the ISEC 2015 study. In addition to being the center of operational and transitional testing, it is likely that the Marine Node will see training on its list of functional responsibilities.

4.4 Marine Node Culminating Demonstrations

4.4.1 MN #1 Tether Terminus

This demonstration will show that the tether can remain attached to the reel in/reel out housing with the IOC number of climbers attached. The reel in/reel out function will operate at IOC speeds to help manage tension and avoid space debris.

4.4.2 MN #2 Position Management

This demonstration will show that the FOP can maintain its position to the IOC accuracies in the presence of wind, current, and tension. The demonstration will include movement of the platform at IOC speeds to avoid space debris.

4.4.3 MN #3 Attach and Detach

This demonstration will show the ability of the climber and the tether to attach and detach climbers to/from the tether

4.4.4 MN #4 Power

This demonstration will show that the FOP can supply power to the climber for the initial ascent through the earth's atmosphere.

4.5 Marine Node Success Criteria

Each subsequent event steps off from previous successes, raises the standards achieved in the previous events and a new event is executed. With each success, technology maturity is attained, engineering approaches are validated and investor's resources are deposited. The test and demonstration sequence for the Marine Node Segment culminates with technologically mature and valid tests. These tests are much more complex as most of them must be conducted in the open Pacific ocean, some even below the surface of the ocean. For the Marine Node Segment, the A&R team foresees four successful Culminating Demonstrations before system engineering and implementation planning can begin in earnest. Success Criteria are not yet identified as information is still needed for a successful result of the demos. The audience at the 2014 ISEC Conference also advised us as to what they thought were the criteria for success. Standards are generally set at the entry to the next test or demonstration event. Criteria are to standards, as weight is to one hundred pounds. The attendees advised that the Marine Node needed to be successful via these criteria:
- Security
- Seismology
- Physical
- Clearances
- Sea Life effects
- Environmental effects
- Keep out zone limits
- Crew safety
- Communications connectivity
- Positioning/repositioning

The totality of the technological maturity and engineering validation activities creates a major test data archive as well as an extensive data processing and performance evaluation repository. The repository of results will be used in determining if the test or demonstration satisfied the success criteria's standards.

4.5.1 MN #1 Tether Terminus

This demonstration will show that the tether can remain attached to the reel in/reel out housing with the IOC number of climbers attached. The reel in/reel out function will operate at IOC speeds to help manage tension and avoid space debris.

4.5.2 MN #2 Position Management

The FOP can maintain its position to IOC accuracies in the presence of wind, current, and tension. The demonstration will include movement of the platform.

4.5.3 MN #3 Attach and Detach

The climber and the tether are able to attach and detach climbers to/from the tether.

4.5.4 MN #4 Power

The FOP can supply power to the climber for the initial ascent through the earth's atmosphere.

4.6 Marine Node Summary

The Marine Node Segment will contain the most mature technologies inside the total transportation infrastructure. However, the one exception is that the tether terminus subsystem must be designed from TRLs in the range of 3 to 5. This is part of the strength of the design as the initial segment of the space elevator architecture must be perceived as customer friendly and operator compatible. The testing leading up to the culminating Demonstrations can most likely be accomplished close to design/construction locations. However, the success oriented test sequence must lead to the final demonstration in the Eastern Pacific ensuring readiness for receiving the end of the tether as it is lowered to the ocean surface.

5 Tether Segment Roadmap

5.1 Introduction

The concept is simple; a one-meter wide tether going from the surface of the Pacific to the Apex Anchor at 100,000 km altitude. Figure 5-1 illustrates two Marine Node Earth termini for a pair of tethers: the basis for a Space Elevator Initial Operational Capability.

Table 5-1 ISEC's Space Elevator IOC Architecture [chasedesignstudios.com]

This chapter lays out the Tether Segment with Figure 5-2 as the Tether Segment roadmap. The Tether Segment's path to Space Elevator IOC is seen as resolving its four primary functional manifestations.
- Development of a Tether that is Long and Strong
  - Long is 100,000 kilometers
  - Strong is able to support the 100, 000 Km tether itself and several active climbers.
- Tether Control
  - Move the tether when needed
  - Dampen movement when needed
- Tether Motion Simulation and operations modeling
  - Predict the need for movement due to multiple factors

- Tether Deployment including the deployment of 100,000 Km of Tether to Space in a manner that enables ground attachment – at the Marine Node

The Tether is the ribbon or 'elevator cable' that the Tether Climber will grip as it rises toward its destination. The Tether must be strong enough to support the weight of the Tether itself, the climber(s) clinging to the Tether, and their payloads. It must also withstand perturbations due to influences from the moon, sun, and other space environment affectations. As indicated by the roadmap, and for the sake of the above primary roles, an ordered taxonomy of sequenced tests will be conducted. In support of this test activity, a broad definition of the Tether Segment's environments will be sought; gaining an understanding of tether motion dynamics, perturbations from the Sun, Earth, and Moon; climber induced perturbations on the tether; and, the hazards/damage from debris strikes on the tether. Ultimately, the technological and engineering validity of the Tether Segment will be shown in four culminating demonstrations.

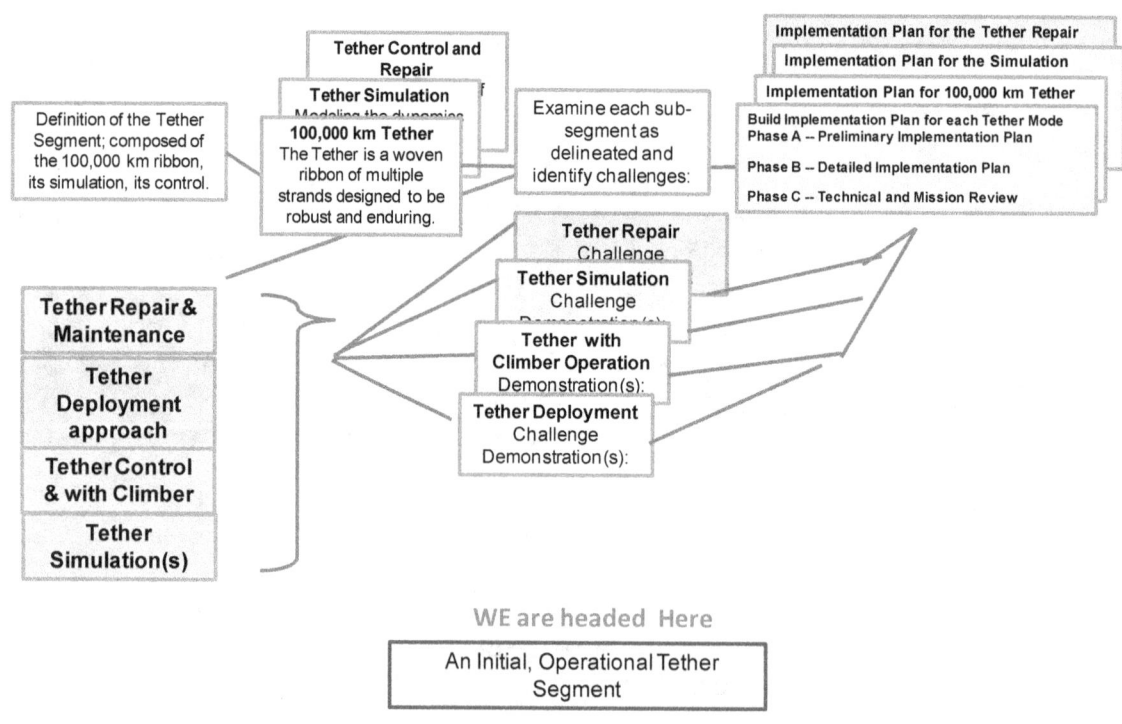

Figure 5-1 Tether Roadmap

5.2   The Definition and Mission of the Tether Segment – The" Ribbon"

This carbon nanotube entity is about one meter wide at the bottom, a few mils thick, with a taper ratio of six, and approximately 100,000 km long. The space elevator will have one end fixed to the Marine Node while extending to an Apex Anchor. Tether/ribbon operations will be conducted mostly by operational personnel at the HQ/POC. Their principle responsibility will be to know the location and expected motion of each element of the space elevator. The requirement is to understand how to adapt the tether's natural motions for operational needs such as Climber motion, initiation of climb, avoidance of space debris, and motion around the GEO node. In addition, the team will monitor the health of the tether and schedule repair functions to be carried out [including any "splicing" that might be necessary for construction of additional tethers]. The strength of the tether determines how many climbers can be on it at any given time.

The mission of the Tether is to provide a basic transportation mechanism for the movement of Climbers and their payloads from the surface of the earth to LEO, GEO, and beyond. The mission includes returning payloads to disposal orbits or to the Earth.

5.2.1 Tether Segment – Node to Node association

Operations of the Tether are controlled at the HQ/POC. The Apex Anchor Node and the Marine Node respond to control instructions resulting in the Tether being reeled in or out at the Apex Anchor or the Marine Node. These reel actions support center of mass management and mission oriented movement. Tether motion for avoidance of space debris would be initiated from one or both of the Nodes. In addition, the Nodes will be reeling the Tether in and out as required for various tasks such as damping tether end vibrations and reacting to emergencies.

5.3 Tether Segment Pathway

The Tether Pathway is shown in Figure 5-2 below. The roadmap team started with these definition and mission statements, and sought to portray how to get to the Segment's planning phase; eventually building the System Engineering based implementation plans in three steps; with depth of detail increasing in each planning step. This segment's vertical disbursement – 100,000 kilometers – puts a significant twist to the test and demonstration portrayal.

The A & R team identified a number of technology and engineering challenges that need to be resolved in order to get to the planning phase. Once again, there were technology needs to be matured and engineering approaches to be validated. The maturation and validation efforts will again be manifested in a sequence of tests, experiments, inspections, and simulations as they grow in thoroughness and complexity. As the Tether roadmap portrays, tests in some four major arenas need to be executed. A test and demonstration taxonomy [test campaign], for each of the arenas is needed:

- Development of a Tether that is Long and Strong
    - The Tether Pathway must include traversing experiments and tests to validate that the Tether is: Long and Strong.
- Tether Control

- Tether Motion Simulation and operations modeling
    - Predict the need for movement due a variety of number of factors
- Tether Deployment competence so that the deployment of 100,000 Km of Tether in Space enables IOC.

Obviously, the test taxonomy execution can "share" the results across the functions; but, it is likely that different Tether altitudes configurations will require different engineering validation efforts within the test and demonstration taxonomic execution. The point is that Tether performance will vary for LEO versus MEO versus GEO and those destined beyond GEO.

The Tether Segment and the set of needed tests, inspections and examinations it needs to reach readiness prior to design development, is exactly what the entire Space Elevator's future is all about. The Tether Segment is composed of more unknown unknowns than the rest of the Architecture. An extensive campaign of tests, experiments, inspections and analyses is needed.

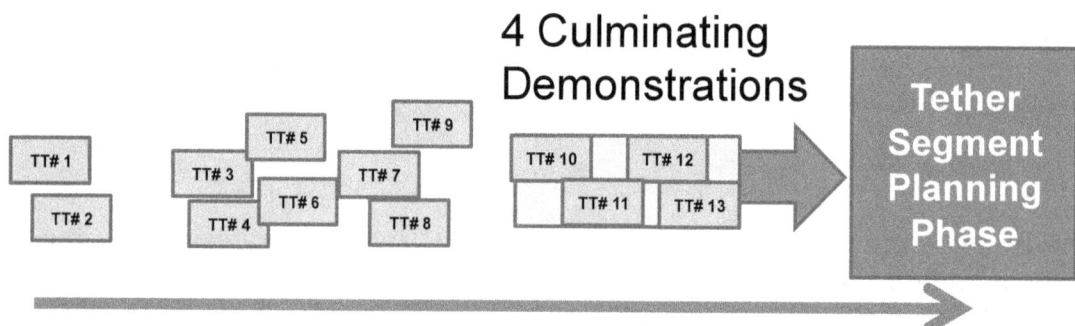

Figure 5-2 Tether Segment Pathway

The totality of the technology maturity and engineering validation activities creates a major test data archive. The repository of results will be the used in determining if the test satisfied the success criteria's standards. With each success, technology maturity is attained, engineering approaches are validated and investor's resources are deposited. This is shown in Table 5-1 below.

#	Term	Demonstration	#	Term	Demonstration
TT #1	near	Evaluate the Preliminary Tether dynamic model	TT #8	mid	Conduct preliminary response tests to reel in / reel out by Nodes
TT #2	near	Tensile strength testing suite begun	TT #9	mid & far	Conduct multiple sever characterization tests. Failure mode analysis.
TT #3	mid	Conduct preliminary center of mass evaluations	TT #10	near	**Culminating Demonstration** – The Tether dynamic simulation
TT #4	mid	Based on the Tether Material LRIP, examine tensile strength and friction / grip tests	TT #11	far	**Culminating Demonstration** – Tether Deployment challenge
TT #5	mid	Examine perturbation "active damping" (thrusters) effectiveness	TT #12	far	**Culminating Demonstration** – Climber capable Tether Operations demonstration
TT#6	mid	Produce Tether repair samples and conduct strength and grip tests	TT #13	far	**Culminating Demonstration** – Tether Repair demonstration
TT #7	mid	Conduct dynamics modeling validation tests with a pathfinder			

Table 5-2 Tether Culminating Demonstrations

The test and demonstration sequence for the Tether Segment culminates when the Tether Segment is able to show technologically mature and engineering valid tests, such as:
- Provides validated high fidelity model of dynamics with atmosphere, climbers, FOP, GEO Node.
- Support IOC Strength requirement (strong enough to allow for climber(s) with payload operations).
- Deploy from a High Space location and captured by the Marine Node Tether Terminus.
- Be maintained and repaired robotically (including recovery from a low altitude sever condition).
- Support IOC Life requirement.

5.4 Culminating Demonstrations

For the Tether Segment, the A&R team foresees four Culminating Demonstrations. These must be completed satisfactorily before implementation planning can begin. Specifically, four separate grand challenges must be resolved before the Tether Segment can be adjudged as ready to be built

As the Tether Segment test campaigns proceed, an empirical technology survey is being assembled composed of the extensive test data generated with each test event and subsequent analysis. This survey is thus the technical framework for each of the culminating demonstrations.

- Tether simulation challenge – The Roadmap and Architectures team recognizes the intimate and dynamic relationship amongst the forces in space near the geosynchronous altitude, near Earth, and due to the moon, the sun and the tether climber aboard the Tether. The challenge is to build a set of models within the simulation of these dynamic interrelationships. In addition, one must establish the models as the basis for the vitally needed operations model used by the HQ/POC to monitor and control the Space Elevator system. The Operations simulations demonstration must be a rigorously accurate link between the Space Elevator and the space environment.

- Tether Deployment challenge – The straightforward idea of dropping the Tether from a satellite to the Marine Node for initial deployment is a lot more involved than it initially appears. The tether first released will likely be a thinner, lighter version of the operational ribbon. This process is the notion of throwing a string across the canyon to pull the rope across to then pull the cable across. A similar sequence is envisioned here. However, the tether must drop through a variety of space environments over a huge distance.

- Climber capable Tether Operations demonstration – This culminating demonstration is the quintessential activity of the Space Elevator. It will show the movement of a climber along the long and strong tether, because it has the grip and power to do so.

- Tether Repair demonstration – The final culminating demonstration is to show how the ribbon can be spliced and repaired

- Pathfinder -- It is pretty clear at this point that a Pathfinder demonstration - or a set of such – will likely be the sensible course for the deployment of a shorter ribbon from a satellite and the movement of the climber along the same ribbon. This concept would be one of the early demonstrations lending confidence that the four culminating demonstrations would be successful. Preliminary discussions have led to an idea focused upon a preliminary "test flight" of a 1,000 km tether, at 2,500 km altitude center of mass, with deployment from a satellite and with tether climbers. This could be accomplished within the near future as it would not require the operationally capable tether strength, but would require a full understanding (modeling) of a 1,000 km tether demonstration.

5.5 System Model

A high fidelity computer simulation will become the baseline for comparing designs of major segments of the space elevator. As such, it should be initiated as soon as possible so that it may assist the design process from the beginning. It seems to us that we should push for a model, placed inside an active simulator with the ability to "tweak" the inputs to understand the outputs. We see a "gold standard" model that would have the ability to represent and model the following key elements of the space elevator:

- Model Tether tension [along the cable of course, but especially near the Marine Node, Apex Anchor and GEO Node]
- Model the ability to reel-in and reel-out at both the Marine Node and the Apex Anchor [with a growth to understand reeling in and out at the GEO Node]
- Model the effects of the Sun and the Moon, with an understanding of the major factors with respect to the Earth's non-homogeneous nature
- Model Floating Operations Platform motion
- Model the effects from the atmosphere
- Represent elasticity, how much stretch? how resilient? Consider stiffness, torsion and shear
- Allow inputs for control of the dynamic motion [thruster action at Apex Anchor, GEO Node, Marine Node, and individual tether climbers.]
- Model the release of climbers/payloads below GEO, at GEO and above GEO
- Represent the electrical properties of the tether, Including conductivity, capacitance and coupling to the ion and electron densities along the length of the tether
- Model space weather effects under solar storm and quiescent conditions
- Take into account the effects on the tether from solar photons radiation pressure (Compton and Rayleigh scattering) photo-electric effects
- Model the currents and fields induced on the tether, including spacecraft charging effects
- Model the motions induced in the 100,000 km tether by its motion through the electromagnetic fields of the Earth (ionosphere, radiation belts, magnetosphere), and solar wind.
- Accommodate varying masses of the tether, the Apex/GEO Node and up to 7 climbers
- Model special case of sever of cable at several altitudes [most likely 1,400 km]

The modeling tool will have a variety of uses. It will be used early for technological feasibility studies including early design. It then can be used in early development and testing. As the model gains fidelity with incorporation of Operational data, it will be used for mission planning and refinement of Operations procedures including anomaly resolution. The prediction capabilities will enable planning for "what if" scenarios. The model will "morph" as more and more Operational data is incorporated thereby increasing its fidelity.

5.6 Tether Segment Summary

The Tether Segment is the critical technology for the concept of space elevators. Four very significant concerns are: 1) producability of long enough/strong enough material, 2) significant tensile strength material, 3) design for repairability, and 4) long enough life expectancy for commercially viable space elevators. Indeed; as the tether material goes – so goes the space elevator. The culminating Demonstrations illustrate this as;

- The first demonstration is a simulation
- The second is a simple deployment of a material to show "long enough – strong enough,"
- The third leverages another significant achievement – gripping of the tether by climber,
- The last is the concept of repair of the tether over the 100,000 km length

The bottom line is that the tether, and especially the tether material, is the pacing item in the development of the space elevator. One critical capability is the ability to splice and repair the tether in space. As such, much emphasis must be placed on the Tether Segment pathway of demonstrations.

6 Tether Climber Segment Roadmap

6.1 Introduction

This section delineates the Tether Climber Segment and the path that must be traversed to accumulate the technological and engineering competence to begin the planning process to build the Climber Segment. Another purpose, perhaps the key purpose, of the Climber Segment's pathway toward implementation planning, is that the tests and demonstrations will convince investors that the Climber Segment can be built to match the vision of the Space Enterprise. We know our destination and will follow our roadmap. Tether Climbers have been portrayed in many fashions in the past as many of the assumptions were different in each of the previous architectures. The next two figures (from ChaseDesigns) are representative of designs that reflect the needs of space elevator tether climbers.

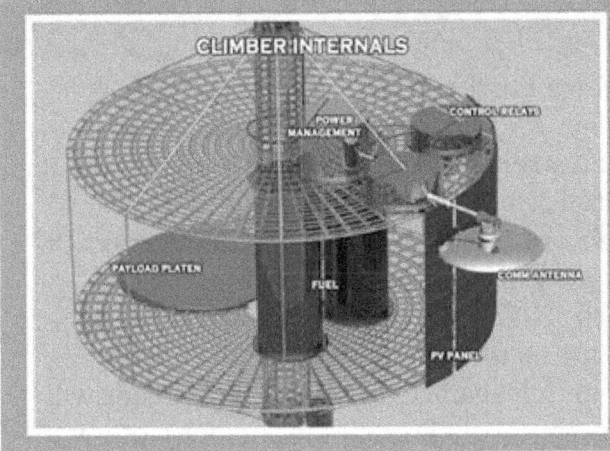

Figure 6-1 Tether Climber Schematic

Figure 6-2 Emerging from Protective Box

6.2 Tether Climber Segment Definition and Mission

The Climber is the entity that ascends along the Tether while carrying a payload. Once outside the atmosphere, the climber is powered by solar cells. The Climber will be fully instrumented and send health, status, and position telemetry to the ground based Telemetry Tracking & Command facility. The Climber will provide power to its satellite payload and can relay payload health and status information. It can climb at rates from meters per second to tens of meters per second. It may have multiple gears and will have a robotic arm. Initial thinking is that it will weigh six metric tons and be able to lift 14 tons of cargo. Some Climbers will carry tether repair apparatus and will execute repairs on the ascent.

The mission of the Climber is simple. It will deliver customer payloads to desired locations along the Tether's path, safely and routinely. Climbers will initially be delivered to the Marine Node as cargo. They will then be mated to the Tether by the Operations team and loaded with customer payloads. The Climber will then proceed under the control of the Operations center at HQ/POC. The current vision of the Climber is cylindrically shaped enabling significant room for large and oddly shaped cargo.

There are many challenges facing the Tether Climber Segment. The essence of the space elevator is delivery of payloads to space cheaply. The following challenges have been identified.

- Climber structure design and production: There are two competing technology trends going at this time: CNT materials making spacecraft structures and batteries lighter and stronger. And, a change in requirements as there will be NO shake/rattle/roll stressing forces from launches.
- Climber/Tether gripping approach and mechanisms: the ability to grasp the tether from the Tether Climber is a mystery at the present time. The coefficient of friction, the fragility of CNTs, the gripping mechanism design, and the forces on tether climbers are all to be determined in the future.
- Climber motor: The ability to lift oneself against gravity at great speeds will be a challenge to designers. The current belief is that the climber motor will be tested by very strenuous methods as the whole safety for crew depends upon the motor working for over 200,000 kms.
- Climber Breaks: The ability to stop on the tether for an overnight stay is a requirement. As such, the tether climber must be able to stop at any point in the climb to GEO and beyond. The brakes must be used hundreds of times and need a robust design for reliability. In addition, the return trip to the surface requires very good breaks with a huge thermal load dissipation of energy.

- Climber power for the trip up to GEO: One reality that must be accepted is that the power must come from external sources as lifting the stored power directly against gravity is a non-starter. As such, power source could be RF, laser or solar. Each carries with it tremendous problems and benefits.
- Protective box below 40 (TBR) kilometers: The complexity of working with solar arrays in the atmosphere has convinced many people that the problem was unsolvable. The funny thing is that the concept is so simple, it outsmarted itself. A simple box that lifts itself up to 40 kms can break the Earth's hold on the tether climber.
- Support to Payload [protection, power, communications, etc.]: The customer wants three types of support. Stability of the space elevator and low cost alternatives are also two types of concern.

The Tether Segment is shown in Figure 6-3 below.

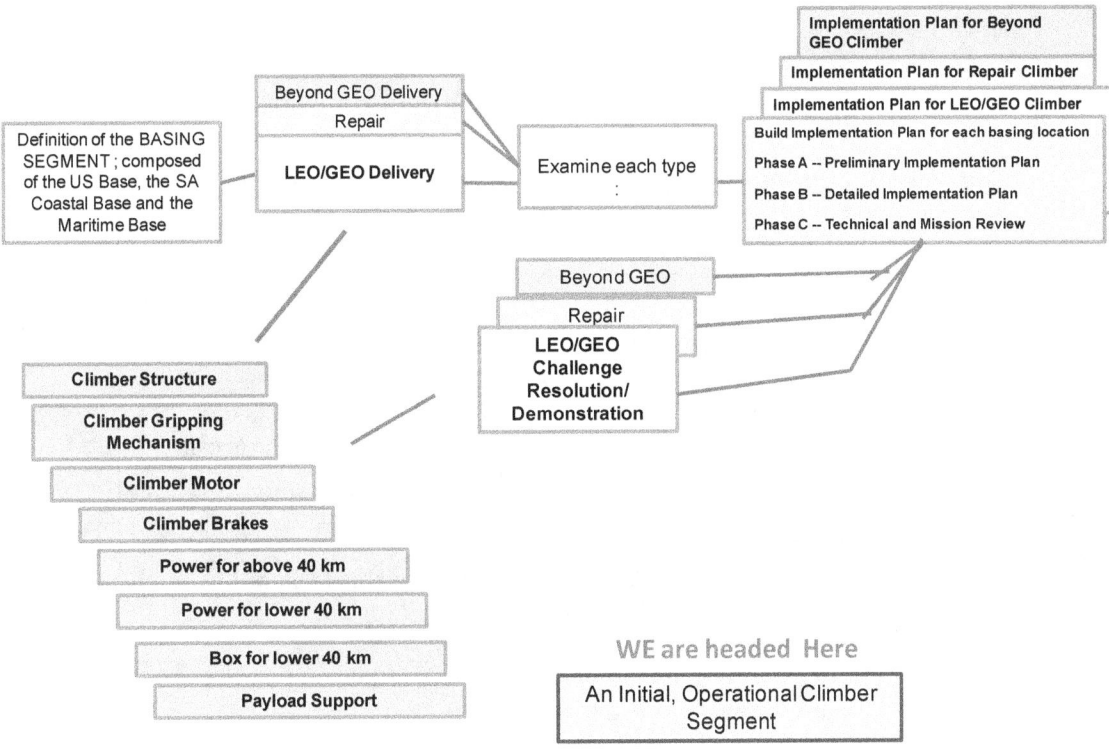

Figure 6-3 Climber Segment Roadmap

6.3   Tether Climber Pathway

The Tether Pathway is shown in Figure 6-4 below. The roadmap team started with this definition and mission. We sought to portray how to get to the segment's planning phase while eventually building implementation plans in steps with depth of detail increasing in each planning step.

41

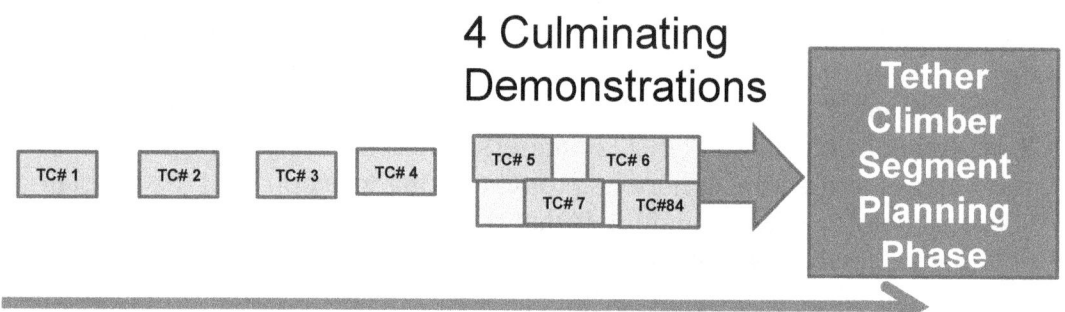

Figure 6-4 The Tether Climber Pathway

The roadmapping team identified a number of technology and engineering challenges that need to be resolved in order to get to the planning phase. As the Climber roadmap portrays some eight major arenas need to be executed. A test and demonstration taxonomy, a test campaign, for each of eight arenas is needed:

#	Term	Demonstration	Dependency
TR1	near	The Climber can provide power for IOC Climber mass and payload above 40km	none
TR-2	Mid	The Climber can provide power for IOC Climber mass and payload below 40km	TBD
TR-3	mid	Climber can survive environment up to 40km	Actual environment needs capable tether. Analysis can buy down risk
TR-4	near	Climber can provide SWAP and data interface	None
TR-5	far	**Culminating Demo:** The Climber is able to climb at IOC speeds carrying IOC payload	Needs capable tether
TR-6	far	**Culminating Demo:** The Climber can reduce speed to safe level and can park	Needs partial length tether
TR-7	far	**Culminating Demo:** The Climber can withstand the earth and space environments while carrying the IOC payload	Needs capable tether
TR-8	far	**Culminating Demo:** The Climber is able to stay on the tether when moving at IOC speeds and parked	Needs partial length tether

Table 6-1 Tether Climber Demonstrations

Obviously, the test taxonomy execution can "share" results across the functions; but, it is likely that different climber configurations will require different engineering validation efforts within the test and demonstration taxonomic execution. The point is that climber configurations could vary for LEO destinations versus tether climbers going beyond GEO.

6.4 Tether Climber Culminating Demonstrations

To move toward these plans, each segment must "demonstrate" that the Climber Segment holds the necessary mature technologies and validated engineering approaches for the design and development activity within the planning phase. Over the course of time, leading to the culminating demonstrations, the Space Elevator team will show that the Climber can do what it needs to do while buttressed by the test data gathered along the way. The eight culminating Demonstrations for the Tether Climber are shown in Table 6-1 above.

- Attach and detach to the tether at the FOP and at the GEO node
- When attached there is no slippage
- Climb the tether at IOC speeds while in continuous power mode with IOC Climber mass and cargo capacity (includes steering to stay centered)
- Generate sufficient power for the IOC motor with TBD mechanism for the first 40 km and solar arrays above 40 km, including deployment
- The Climber is able to:
  - Deploy GEO satellites
  - Deploy LEO/MEO satellites
  - Descend safely at IOC speeds, includes steering to stay centered

6.4.1 Additional Demonstrations
- Beyond GEO: An interesting aspect of this challenge is that the force disturbing the status quo is pointing away from the Earth and modest vs. the gravitational force it has escaped. As such, the culminating demonstration will probably be conducted in the LEO region, but the forces would be the opposite of what is expected. The demonstration would have to be towards Earth, as if it was falling away from GEO.
- Repair of the Tether: Everyone knows that there will be wear and tear of the tether with possible small holes from space debris. As such, a methodology and engineering solution must be developed and demonstrated on how to repair the tether climber. The financial projections show an estimated seven year replacement schedule for a full tether. If that length of time can be accomplished, the profit should flow. If that timeframe is too long, the maintenance might be an alternate approach to new tether segment replacement. Many approaches have been proposed; but, the decision will have to wait until a better solution occurs.
- LEO/GEO Challenge Demonstration: There seems to be a push to have an in-orbit test for a short representation of tether and tether climber. This grand challenge will ensure that everyone is satisfied with the design and results.

6.5 Success Criteria

The development of Success Criteria is critical, as shown already. As such, several criteria must be demonstrated at sufficient levels of performance:
- Attach and detach to the tether at the FOP and at the GEO node
- When attached there is no slippage
- Climb the tether at IOC speeds while in continuous power mode with IOC Climber mass and cargo capacity (includes steering to stay centered)
- Generate sufficient power for the IOC motor with TBD mechanism for the first 40 km and solar arrays above 40 km, including deployment
- The Climber is able to:
    - Deploy GEO satellites
    - Deploy LEO/MEO satellites
    - Descend safely at IOC speeds, includes steering to stay centered

6.6 Tether Climber Segment Summary

The Tether Climber Segment seems to be the most straightforward as most of the tasks have been accomplished within our space community over the last few years. The challenges are more on "how to" rather than inventing something new for an impossible task. However, it is really critical that we approach each segment as a difficult challenge, while confident that the grand challenges will be accomplished.

7 Apex Anchor Segment Roadmap

7.6 Introduction

The dynamics of a space elevator shows that a tether of 150,000 km is in balance without requiring a counterweight. The centripetal force of the part beyond geosynchronous orbit offsets the gravitational forces that are nearer to Earth. In addition, the angular momentum of the upper mass beyond GEO must be in sync with the angular momentum aspects of the lower mass and the orbital angular momentum. A counterweight can be used to shorten the overall length, and thus, a trade-off in overall mass. The recent architectures have standardized at a 100,000 km altitude for the length of the tether with an Apex Anchor of significant mass. The breakout has been shown to be:

 Tether Mass: 6,300,000 kg
 Apex Anchor Mass: 1,900,000 kg

This will result in an Apex Anchor that will have significant tension in the ribbon to adapt to the various forces on the tether dynamics, including tidal forces and tether climber motion. This will ensure that all forces are accounted for and allowed [Earth's gravity, centripetal force, lunar gravity, solar gravity, tether climber forces, and others such as electromagnetic interactions]. This leads to an Apex Anchor that will ensure an appropriate tension in the tether to create extra rigidity and stiffness along the path of the tether climbers.

Mass for the counterweight could be made available from many sources to include near-by asteroids, tether climbers, dead GEO satellites, and additional mass from Earth as required. The initial masses could come from the vehicles used to place the initial tether fibers into orbit and from the small tether climbers used in the tether's construction. Using this material as a counterweight lowers the total mass that has to be lifted into orbit. The choice of capturing an asteroid was an early idea and might become real with a robust space infrastructure as the space elevator becomes routine; however, near-term solutions must dominate. The basic systems engineering requirements for mass of the Apex Anchor are that the mass be easily transported on the tether and not be explosive. Mass is good in the space elevator context.

7.2 Apex Anchor Segment Definition and Mission[5]

However, a key to the development and operations of a space elevator is that the Apex Anchor must be a major element in the operations of the total infrastructure. The Apex Anchor must have the ability to provide initial stability to the tether, maintain stability for

---

[5] Swan, Peter, David Raitt, Cathy Swan, Robert Penny, John Knapman. *Space Elevators: An Assessment of Technological Feasibility and the Way Forward*, IAA, Paris, 2013. Much of the information in this section is paraphrased from the section on Deployment Satellite.

the total space elevator during operations, provide forces [through thrusting] to counter unexpected or planned motion damping, be refueled, coordinate with the Tether Operations Center to dampen harmonic motion and stimulate desired movement of the space elevator tether. In addition, the smart Apex Anchor must work with customers for deployment of mission payloads, as well as capture returning payloads, while refueling and assembling customer's space systems. Indeed, the Apex Anchor will do far more than "just be a mass at the end of the space elevator." As a result of all these demands on the Apex Anchor, the mission is to:

Provide stability for the space elevator and to support customers.

The current design approach has the Apex Anchor growing from the initial space elevator deployment satellite. The process is simple;

- Deployment satellite, a massive space system, assembled in LEO
- Moves from LEO to GEO by efficient rockets,
- Initial deployment of tether; the deployment spacecraft would deploy one end of the seed tether in the downward direction towards the surface of the ocean while raising the massive deployment satellite in the opposite direction, keeping the whole system center of mass at the allocated GEO node. [thrusting required to compensate for angular momentum losses]
- Buildup of the tether commences as small climbers raise and enhance the tether from one that is a seed tether to one that is operationally capable at IOC.
- At the end of the deployment phase, the deployment spacecraft becomes the principle part of the Apex Anchor. This would include computational capability, thruster ability, fuel storage [with refueling ability], communications links to HQPOC, Marine Node, and satellites on the tether. In addition, it must support customer needs for release of payloads to mission orbits or capture and return of payloads to the space elevator infrastructure.

A mental image would show an 80,000 lb deployment satellite that would have a large deployment drum and many – many fuel tanks. The deployment spacecraft consists of:[6]

- "Tether Payload: The tether must be delivered to the GEO node so that it may be reeled out in a smooth and controlled manner. The current assumption is that the whole deployment satellite will become the Apex Anchor as the tether is reeled downward with a commensurate motion upward from angular momentum and center of mass adjustment from the motion of separation. Once the tether is secured at the Marine Node, the deployment satellite would then reel out more of the tether, allowing the heavy mass to reach outward and add more tension, thus stabilizing the tether as well as adding structural features. This initial "seed" tether would have a minimum diameter but still have the required strength to weight ratio to ensure survival while minimizing weight.

---

[6]Most of this section is information and phrases directly from: Swan, et.al. Space Elevators: An Assessment of the Technological Feasibility and the Way Forward., Virginia Edition Publishing, 2014.

- "Reel & Drum: Early in the design of the first space elevator a mechanical system for deploying the tether would be tested extensively. Some keys will be that the speed should be high enough to make the deployment time reasonable [less than 1,000rpm], but slow enough to ensure that the tension is controlled and the dynamics of the deployment are not "out of spec." A spool of approximately 6m in length with a diameter of 2m would provide sufficient mechanical leverage and hold the required initial tether length of greater than 80,000km.

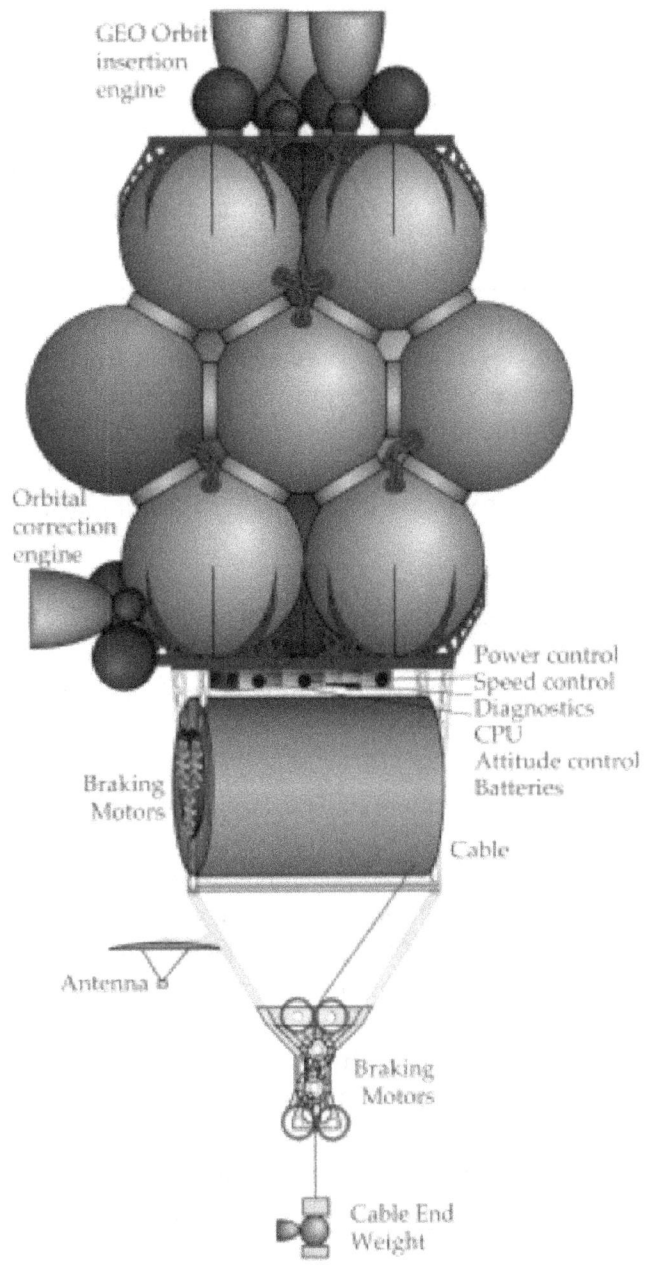

Figure 7-1, Deployment Satellite

- "Structural Elements: The image of the deployment satellite seems to show a whole series of propellant tanks with a huge reel/drum of tether material and a mechanism to release the tether downward. All of these elements must be firmly held together to enable stabilization of the structure during thruster firing and deployment operations. As there is a need to assemble major components of the deployment satellite in LEO, a "smart" design and different approach for a future deployment system could be employed. The components of the satellite that are launched must survive the rock and roll of the first 500 kilometers, but then reside in zero-g. The resulting deployment satellite that must be moved to GEO will only experience small acceleration loads, even using large ion engines.

- "Power Subsystem: As the deployment satellite will require power to operate a space system, the natural element for power generation is solar cells with efficiencies approaching 48% [in labs today] and extremely light for weight savings.

47

- "Attitude Control with Propellant: Attitude control will be essential during assembly and transportation to GEO. While the deployment satellite is in LEO (on its way to GEO), as well as early on at GEO, the attitude control will probably be achieved by spinning masses [e.g., control moment gyros (CMGs] and torque rods [more effective in LEO]. Once the tether has deployed a sizable distance, the gravity gradient factor will help stabilize the deployment satellite.

- "Command & Control Communications: C&C communications will be achieved through relay satellites, as there is a requirement for constant connectivity. As a result, there will be a geosynchronous communications satellite antenna required to track the LEO grouping of sub-satellites, monitor the assembly, and then track the deployment satellite as it goes from LEO to GEO. Once in GEO, the space elevator's communications architecture will kick in with connection to Headquarters from GEO and control of the link monitored continuously.

- "Thermal and S/C Support: Overall support of the spacecraft will involve many disciplines including thermal, radiation, electromagnetic, orbital location knowledge and projection, pointing and stability. During the 14 orbits per day while in LEO, the thermal stresses will be greatest and will manifest themselves inside all aspects of a spacecraft.

- "Thruster Elements: The large task ahead for thrusters is to raise a very large mass from LEO to GEO in an efficient manner without time being a large factor. As such, the cluster of large [1 meter diameter] ion engines will provide continuous thrusting to raise the altitude in a spiral orbit. The specifics of the thrust, efficiency of the engine, and the time it will take to move the mass to GEO will be understood in greater detail as the mission approaches.

Now that the Apex Anchor Segment has been grown from the Deployment Satellite, the actual challenges and testing requirements can be discussed. The major challenges are: Stabilize the initial delivery of the space elevator, establish desired dynamics of the tether, and support customer activities. These are shown in the Roadmap figure shown next. Inside this chart, the challenges are shown and an initial start of a list is provided. These are:

- System Mass Management: the Apex Anchor's principle task is to ensure that the space elevator is stable and continuously providing knowledge of each element of the infrastructure. Critical functions include management of the center of mass of the system as well as understanding the change in mass center or motion as the tether climbers are added and moved.

- Tether Reel-in and Reel-out Capability: This task is relatively simple for the Apex Anchor as it has just released 100,000 km of tether during deployment. The upper end of the space elevator must be able to assist the Marine Node in understanding, recognition of "truth" and control of every element of the infrastructure. The two segments will work together by reeling-in and reeling-out the tether as needed to ensure stability of the system.

- Thruster Magnitude and Direction: During operations, the motion of the Apex Anchor must be controlled. This will require thrusters and the ability to point them in almost any direction with a variable thrust.

- Customer Payload Stabilization, Release and Capture: Once the space elevator initiates customer support, the desires will vary. Three approaches seem to be in demand by potential customers: stable location for mission [stay attached to tether], release to mission trajectory, and capture for placement on a return tether.

These challenges are shown in the Apex Anchor Segment Roadmap Figure 7-1.

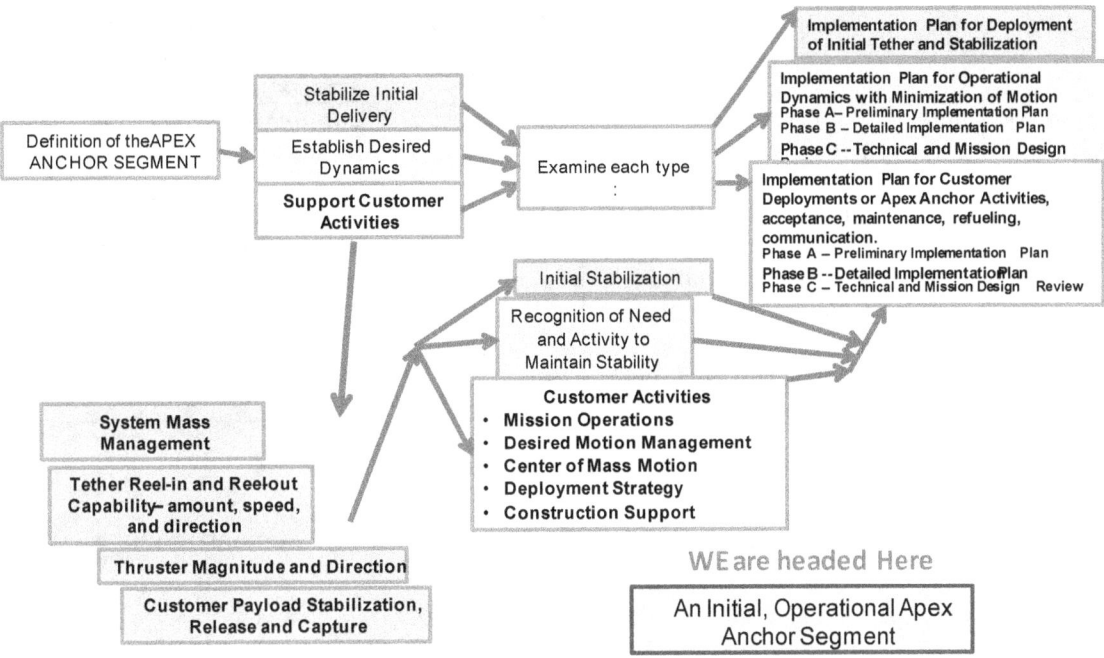

Figure 7-2 Roadmap for Apex Anchor Segment

7.3 Apex Anchor Segment Pathway

The Apex Anchor Segment has many steps to go through for reduction of technological risks and systems integration. This will be across the board; but, each is an essential step towards the fulfillment of the potential for an Apex Anchor.

The following table is a layout of potential demonstrations that will lead to the three culminating demonstrations.

#	Term	Demonstration	#	Term	Demonstration
AA-1	mid	Reel-out tether	AA-6	mid	Design flexibility for customer construction and assembly
AA-2	mid	Reel-in tether	AA-7	far	**Culminating Demo:** Recover from anomalies beyond expectations
AA-3	mid	Stabilize the initial tether situation – center of mass control	AA-8	far	**Culminating Demo:** Customer payload release and capture
AA-4	near	Simulate normal modes [no climbers], deployment, multiple climbers, solar/lunar effects.	AA-9	far	**Culminating Demo:** Customer payload and space tug refueling
AA-5	near	Illustrate thrust profiles to enable stabilization			

Table 7-1 Apex Anchor Demonstrations

Figure 7-3 below shows the layout to reach the three culminating demonstrations.

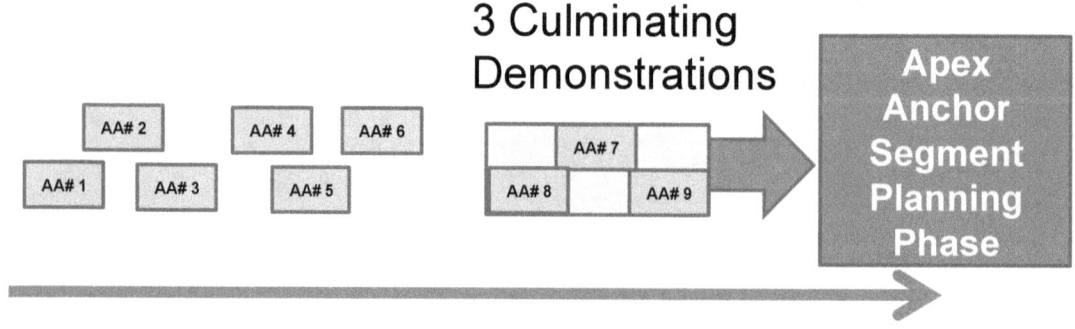

Figure 7-3 The Apex Anchor Pathway

7.4 Apex Anchor Segment Culminating

The three culminating demonstrations for the Apex Anchor are as follows:

- Tether reel in and reel out.
- Thruster magnitude and direction.
- Support Customer Activities: When the customer starts to leverage the space elevator, the Apex Anchor will play a major part as a customer satisfaction arena. As the customer's satellite approaches the Apex Anchor, their needs will range from help in releasing the satellite from the tether, assembly of multiple parts to build a larger spacecraft, and acceptance of incoming payloads to the space elevator to be sent back to the GEO Node or the surface of the Earth. This ability to off-load and attach customer payloads at the Apex Anchor must be demonstrated, probably in space with appropriate accelerations.

7.5 Success Criteria

For the Apex Anchor Segment to be prepared for the Initial Operational Capability deadline, these three culminating demonstrations must be successfully completed. Some major thoughts reference the Apex Anchor Segment are discussed as:

- Establish a dynamically stable tether situation during deployment while reeling-out tether
- Stabilize the total space elevator upon reaching 100,000 km altitude
- Prepare customer payload and then release from the tether to its mission trajectory
- Accept incoming customer payload, attach to tether climber and send Earthward
- Operate Apex Anchor with communications node, dynamic planning and customer support

7.6 Apex Anchor Segment Summary

The Apex Anchor Segment is a very dynamic part of the space elevator. The ability to understand the natural motion and then ensure stability will be its most significant task; however, support to space elevator customers will be paramount in ensuring commercial success. This would include the ability to understand the whole arena of dynamic stimuli such as tether climbers, reeling-in/out of tether, forces from winds, movement of the Marine Node, and stability supplied by a large mass at the GEO Node.

The bottom line for the Apex Anchor is that it will be an intelligent component to the space infrastructure that is called the Space Elevator.

8 Headquarters and Principle Operations Center Segment Roadmap[7]

8.1 HQ/POC Segment Introduction

The HQ/POC will host key elements of conducting the business of transporting payloads to and from space. The business side will be the Enterprise Operations Center while the day to day execution of activities will be segmented out to the various operations centers co-located within the HQ/POC. The HQ will represent the corporation while the POC will consolidate the operational functions of the system of systems. The charts and text below depict a high level view of the HQ/POC.

8.2 HQ/POC Segment Definition and Mission

The following charts and text show the layout of the facility with the table showing significant functions to be handled by the HQ/POC. A characteristic of this is that the operations for the space elevator transportation infrastructure are segmented into multiple operations centers. This co-location is by design as two factors will dominate: (1) the communications architecture will allow 24/7/365 connectivity to anywhere in the infrastructure and (2) co-locations should maximize efficiencies and minimize staffing demands. These operational activities will be conducted remotely, such as at the Marine Node, or in the future at the GEO Node.

---

[7] This chapter borrows heavily from the Space Elevator Concept of Operations, ISEC Position Paper #2012-1

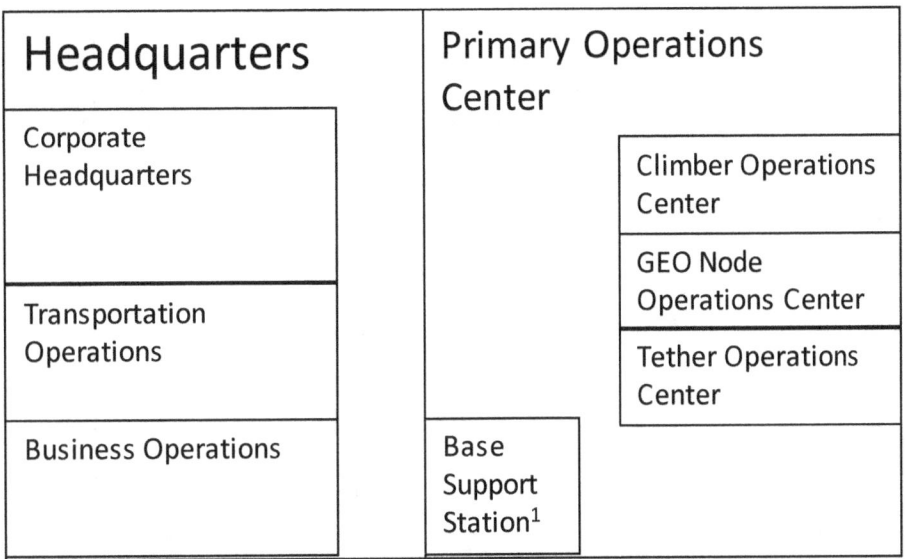

Figure 8-1 Operations Centers

The functions to be accomplished at the HQ/POC, in addition to at the co-located operations centers, are:
- Mission planning
- Activity planning and development
- Mission control
- Data transport and delivery
- Navigation planning and analysis
- Spacecraft planning and analysis
- Payload planning and analysis
- Payload data processing
- Archiving and maintaining the mission database
- Systems engineering, integration and test
- Computers and communications support
- Developing and maintaining software
- Managing mission operations
- Financial management

Headquarters and Primary Operations		Headquarters			Primary Operations Center			could be co-located
		Corporate HQ	Transportation Operations Center	Enterprise Operations Center	Climber Operations Center	GEO Node Operations Center	Tether Operations Center	Base Support Station
Mission Planning	MP				x	x	x	
Activity Planning and Development	APD		x	x				x
Mission Control	MC				x	x	x	
Data Transport and Delivery	DTD			x				x
Navigation Planning and Analysis	NPD				x	x	x	
Climber Planning and Analysis	CPA				x			
Payload Planning and Analysis	PPA				x	x		x
Climber Data Processing	CDP				x		x	
Archiving and Maintaining the Mission Database	AMD			x				
Systems Engineering, Integration and Test	SEIT	x		x				
Computers and Communications Support	CCS	x		x				
Developing and Maintaining Software	DMS	x		x				
Managing Mission Operations	MMO			x	x	x	x	
Financial Management	FM$	x		x				

Figure 8-2 Major Functions

The HQ/POC can be located anywhere, but for the initial concept it will be located in the greater San Diego, California area. It will have communications to all the other elements and will have an operations center staffed 24/7/365. Other sites closer to the Marine Node will be studied. Candidate sites must have both an international airport and a port on the Pacific Ocean. Stability of the government and overall security for personnel, equipment, and facilities will also be factors. The following sections discuss the various operations centers within the HQ/POC, address distribution of the operational activities, and look at required staffing.

8.2.1 Enterprise Operations Center

This infrastructure is the home for all the business operational activities as well as the administrative and logistics functions necessary for supporting the operation of the Space Elevator infrastructure. This location will be the for the lead on all financial transactions [Financial Management function] within the corporate infrastructure throughout the various facilities and centers spread around the world. The corporation headquarters is located at the HQ/POC to ensure day-to-day cognizance of the space elevator's business environment. The Enterprise Operation Center will focus on the revenue and expenses for operations across corporations. Operations will range from strategic planning for the enterprise to the research and development needs of future implementations. Day-to-day operations across the enterprise will be looked at; however, the functions will be to support sound business principles while conducting a transportation business. The Enterprise Operations Center may look something like the European Space Agency's Operations center shown below.

Figure 8-3 Notional Operations Center

8.2.2 Base Support Station

If not co-located with the Headquarters, this will be the forward support base for operations. Its focus will be on processing supplies, satellites (climber payloads), and climbers for transportation to/from the Marine Node. This will probably be located at a port for loading purposes. Staffing estimates are included with the HQ/POC estimates.

8.2.3 Transportation Operations Center

All transportation aspects of the enterprise will be controlled from the TOC. This is where payloads and climbers will be tracked. Location and status information will be monitored from the factory to the Business Support Station to the Marine Node, and then followed up the space elevator. Monitoring of returned payloads and climbers will also be done here. Arrangements for ocean going vehicles will also be conducted here as well as planning for helicopters and air transportation.

8.2.4 Climber Operations Center

This is where the majority of tether climber operations are conducted. Many of its activities include plans for delivery and maintenance of the various tether climbers that are on the tether and in process for being attached to the tether, or unattached from the tether. It is likely that the center will keep track of all tether climbers that are raising cargo to altitude, those docked at the GEO node, and those descending. Climber operations will consist mainly of monitoring the health and status of the climber: rate of climb, temperature of the motor(s) and wheels, and other health and status data. The monitoring of climbers will commence upon attachment to the tether. Marine Node personnel will have access to all telemetry being sent to the Climber Operations Center after attaching and before ascent begins. Initiation of climb will be directed by the Climber Operations Center. These operations will leverage satellite operations that have been conducted since the early 60's. Some climbers will perform repair operations, which will likely be a combination of autonomous and operator involved activities. The climber, with payload, will ascend at a rate of meters to tens of meters per second using energy from solar panels. This would enable a round trip of about two weeks, or less. During periods when solar power is not available, the climber will remain stationary. Batteries on the climber and/or the satellite may be used to enable communication with the FOP and to perform housekeeping tasks. Movement of the tether to avoid space debris might also require the climber to park. An additional function to be performed in support of climber operations is the release of instrumented balloons from the FOP to collect high altitude weather information to support the operations of climbers under 40 km.

At the desired altitude, the shroud on the satellite will be removed at the direction of the satellite owner. The climber, likely using its own robotic arms, will assist in removing the protective covering of the satellite and positioning the satellite away from the tether so that any required thrusting will not harm the climber or the tether. Such an arm would also be useful for receiving and securing satellites for return to Earth. One concept for an arm (from Tethers Unlimited) is shown in Figure 8-4 below.

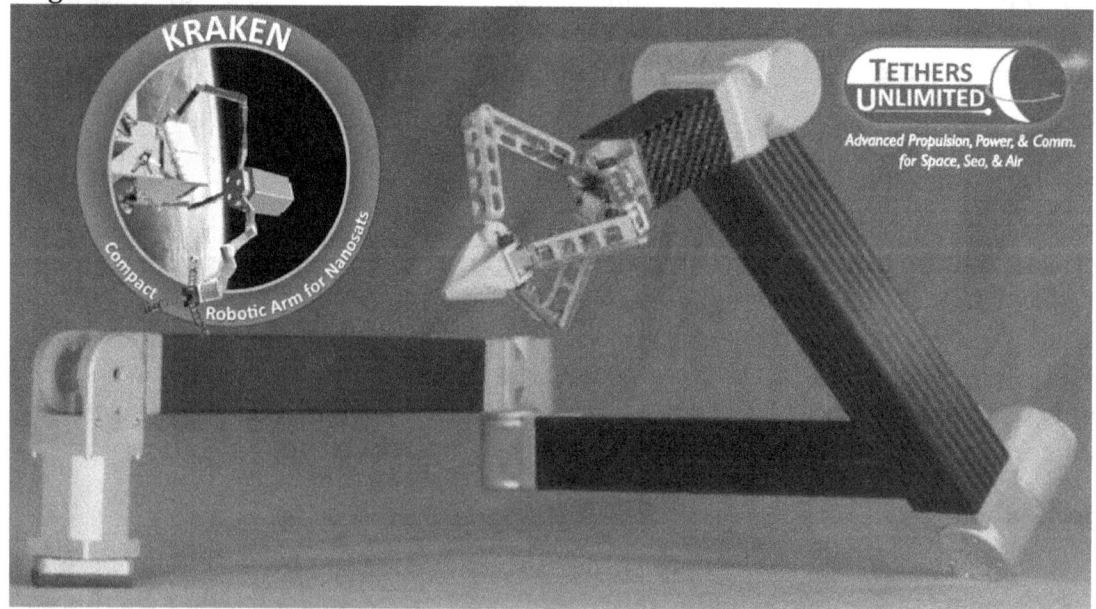

Figure 8-4 Robotic Arm Example

After deployment of LEO satellites, the climber will continue to climb to the GEO node. There, it will pick up any payloads for return to earth and begin its descent. When multiple space elevators are in operation, the climber will be de-mated from the tether and ferried to another GEO node for the descending tether. It may be loaded with a satellite for return to the earth's surface, it may pick up one on its descent, or return empty. At this point, the climber may begin its return trip. In initial operations, the climber may continue to higher altitudes with or without additional payloads and reach altitude where it will act as part of the apex anchor.

As the climber approaches the surface, it could deploy drag chutes to slow it down and brakes to bring it to a stop on the FOP. When arriving at the FOP, the climber will be de-mated for inspection and repair, as necessary. If another satellite payload is ready for lift, it will be mated to the climber and the climber will be mated to the tether. A satellite may also be mated after the climber is mated to the tether. As the second space elevator becomes operational, returned climbers will be transported to the partner FOP.

8.2.5 Tether Operations Center

Knowledge of the three-dimensional location of all elements of a space elevator's tether [assume each element is approximately five km long] is important to the operation of the total system of systems. Each tether climber location will need to be known continuously monitored as to speed and expected location in the near future. This will enable the tether operations crew to understand their situation at all times. Concern for space debris impact becomes critical to successful operations through location maneuvering. This will consist mostly of operations personnel monitoring the probability of space debris impacting a tether. US Strategic Command (USSTRAT) will send out advisories predicting close approaches between large objects and the space elevator tether(s). Tether managers will decide whether to reel in or reel out the tether (and how much) to avoid possible collisions. Reeling out just a few meters of tether from the GEO host can impart tens of kilometers of lateral distance. Looking at an altitude of 660 km:
- 10 meters spooled out from GEO results in a little over 26 km lateral movement in the LEO region
- 100 meters spooled out results in about 83 km
- 1 km spooled out results in about 265 km

Managers will also decide whether or not to re-position after the predicted collision time. Tether operations will include gathering of positional data for the tether(s) and reporting to USSTRAT. Note that debris avoidance can also include re-positioning of the Marine Node terminus itself.

8.2.6 Apex Anchor/GEO Node Operations Center

The initial Apex Anchor/GEO node will have only robotic operations. As such, the operations center will be remoted to the HQ /POC. In the future, when humans operate at the GEO Node, the operations center will be handling all tasks assigned to the large facility in GEO. Those functions include off-loading and on-loading payloads to climbers. Refueling will be a principle mission at the GEO node as it is relatively inexpensive to deliver fuel rather than launch with it.

8.2.7 Organization Chart

Below is a notional organization chart for the HQ/POC.

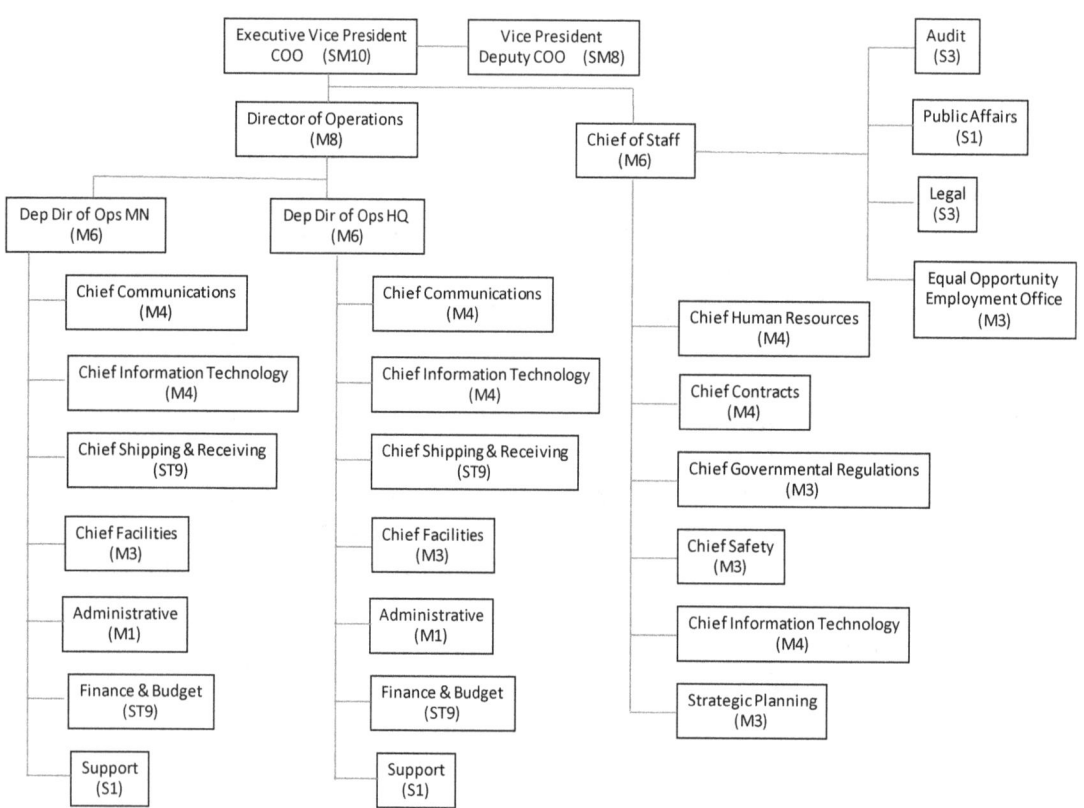

Figure 8-5 HQ/Primary Operations Center Organization Chart

8.2.8 HQ/POC Pathway

As the HQ/POC, at least part of it, will need to be in place early a pathway in the form of those presented in this document is not applicable. HQ/POC functionality will be implemented as program needs dictate in more of a "Build Out" fashion.

The HQ/PC will be housed in an existing structure or one built especially for it. The Business/Support functions will use existing technologies for normal office environments. The Mission Operations centers will also use existing technologies for the functions described earlier. Each of the functions will likely start out modestly and grow into its full up capabilities. For example, the tether operations center may start with a small number of servers and hosts for the applications associated with tether operations. The high fidelity model would be installed and maintained by the developer with increasing numbers of applications and changes to applications. Operations personnel would be involved in the early stages of designing and building the models.

8.2.9 HQ/POC Success Criteria

The HQ/POC will be successful when it can:
- Communicate with every node
- Control the tether
- Control the Marine Node
- Control the Apex Anchor
- Control system center of mass
- Manage system models

8.3 HQ/POC Segment Summary

The HQ/POC Segment can be modeled as the brains of the space elevator system. As such, there will have to be an early, and substantive, version of the operations center to support the culminating demonstrations and systems integration tests. The demonstration pathway will indeed "touch" all the segments as well as ensure that all the internal components excel at their tasking. Many of the early tests will validate the choices of automation or "human-in-the loop." Indeed, the final series of Grand Challenges will be conducted from an early implementation of the Primary Operations Center. During progress towards operations of the system, the developers will have provided applications for each function supporting each segment. One critical role, which usually is underappreciated during development, will be addressed in the business enterprise.

9 System Integration

9.1 Introduction

In the far term, and concurrently during early development, the segments are busy validating that they are able to accomplish what they set out to do. With these successes inside each segment, the Space Elevator development team must begin determining whether the segments can operate together. A series of system integration tests will be added to the test campaigns. These systems level tests will validate the interoperability of the five segments. This leads to the realization that a full range of Integration testing must be conducted concurrently with the Culminating Demonstrations. Essentially, this testing will validate that all the pieces work together. As can be seen by the activity depicted in Figure 9-1, integration testing will be conducted during a very busy time in the maturation of the Space Elevator Architecture. Integration testing is, at the simplest level, a series of giver – receiver validation tests which, when successful, show that one segment can communicate with another segment and get that other segment to do what is needed for compatible operations.

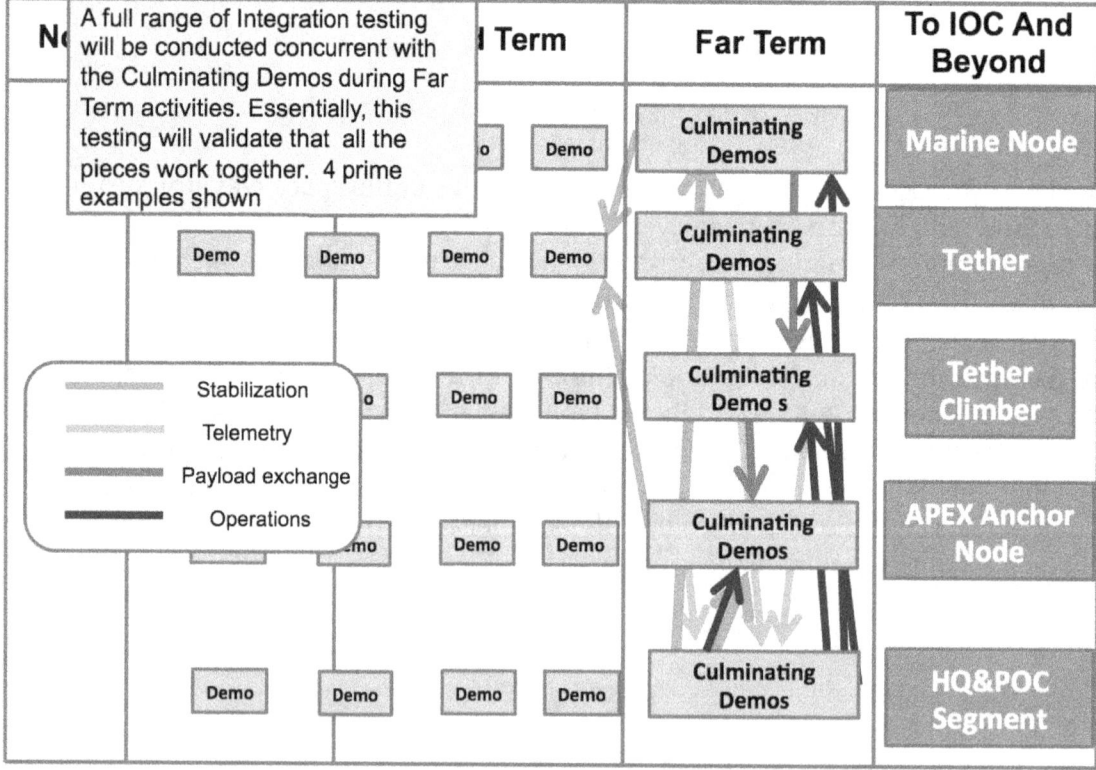

Figure 9-1 Integration Testing-Birthing the Architecture

9.1.1 The Humanity of Engineering Interfaces

As is always necessary for endeavors such as this, the authors asked several colleagues to review the report as written. As was expected, we received a series of citations to fix our grammar and clarity of message. In one case however, a reviewer, Hal Rhoads, captured a spirit of our future that we all feel.

Our section 9 is a fine discussion of the importance of and the complexity of the Space Elevator's integration challenge. Indeed, the integration of the Architecture's pieces may be our biggest engineering challenge. But Hal's feedback was a motivational message. He said:

> ***Our generation learned and relearned (a) frustrating, painful fact of engineering life at enormous cost in time, treasure and occasionally human lives lost as we built the rocket-era space programs. We learned that success grows from free communication, mutual respect, and (believe it or not) friendship within the team. (We must do our) level, professional, passionate best to eliminate barriers and conflicts between individuals and teams. (We should) arrange to have every member of the Space Elevator Enterprise work in the same space, drink coffee from the same pot every morning, and talk things out over beer every week, it would save ... billions of dollars and years of development time.***

The authors stand in humble deference and respect to what Hal said. The architecture we seek is grand, glorious, and beautiful; and may be the basis for mankind's future. But, it can be achieved only by a close knit group, working in respectful unison, on track to that common goal. Now we are ready to begin to get started!
    Thanks Hal.

9.1.2 Interface Management

Successful integration is virtually impossible without a sound Interface Management function in the System Engineering, Integration and Test (usually known as SEIT) organization. As we approach the Requirements Analysis and Requirements Allocation activities preceding the Design Synthesis activities (the standard System Engineering paradigm) we will mandate that the SEIT organization "own" all the system and segment interfaces. The interfaces will be documented in Interface Control Documents, one for each interface, and will address all seven layers of the Open Systems Interconnection Reference Model. Each segment will have a functional requirement to build its half of the interface. An early activity will be to define, via interface control document (ICD), what data is exchanged over each interface. It is anticipated that we will mandate use of open and standard interfaces. We will shun proprietary ones.

9.2 System integration objectives

In the integration testing activity, we foresee four basic objectives. Each objective is achieved when a series of specific segment-to-segment interchange functions occur properly. The four objectives are tether stabilization, telemetry receipt and analysis, payload exchanges, and operations management.

9.2.1 Tether Stabilization.

The first integration objective is Stabilization, depicted in Figure 9-2. This objective is demonstrated by the Marine Node and/or the Apex Anchor Node controlling the Tether (and climbers on it). Control is accomplished by "reeling" the Tether ribbon in and out. These Nodes do so in response to instructions received from the HQ/POC. Clearly, this integration test is the signature event of the Space Elevator Architecture. Failure is not an option. If control of the Tether, the Climber, and its payload is not a normal constant status of the Architecture, then we aren't going there. Obviously, this final integration testing will be preceded by a long series of pre-tests and some space based pathfinder tests.

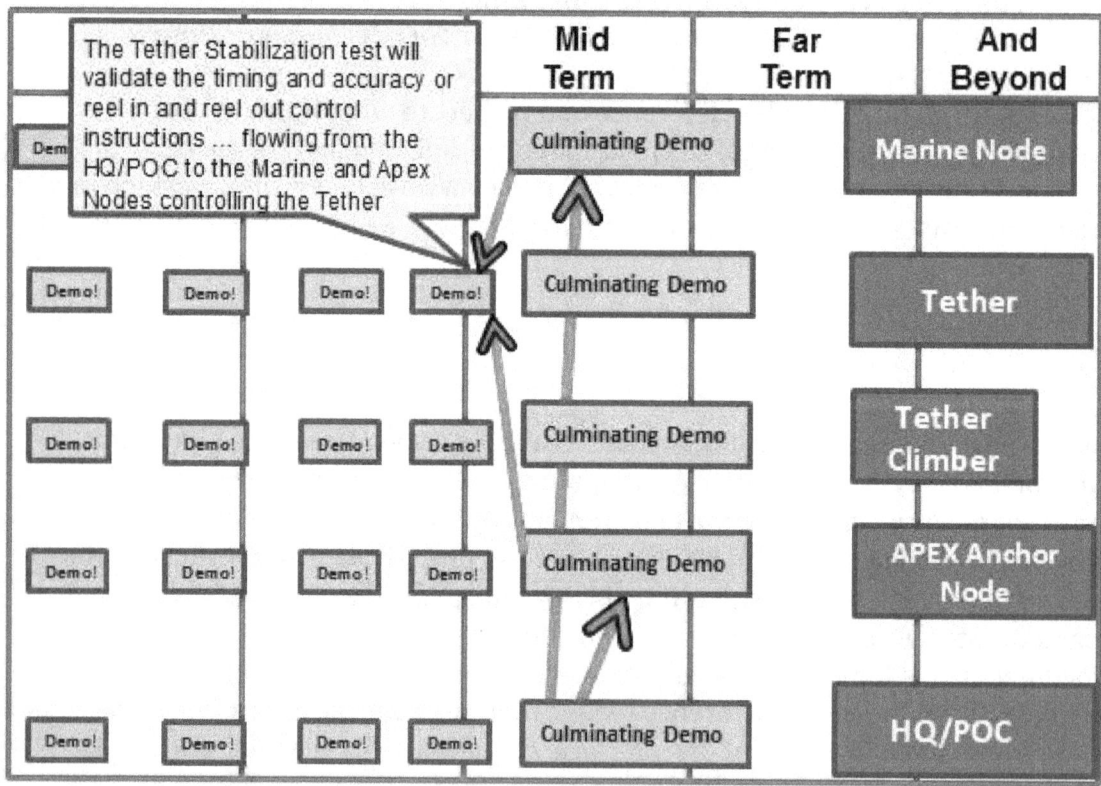

Figure 9-2 Tether Stabilization

9.2.2 Telemetry Receipt and Analysis.

The second integration objective is Telemetry exchange, depicted in Figure 9-3 below. This objective is demonstrated by the HQ/POC sending and receiving the full series of monitoring information from the space based portions of the Architecture and sending a full series of commanding information to the space based portion. Nominally, this type of integration testing is standard fare for those of us in the space business. However, it will not be taken lightly and will be preceded by a long series of pre-tests and some space based pathfinder tests.

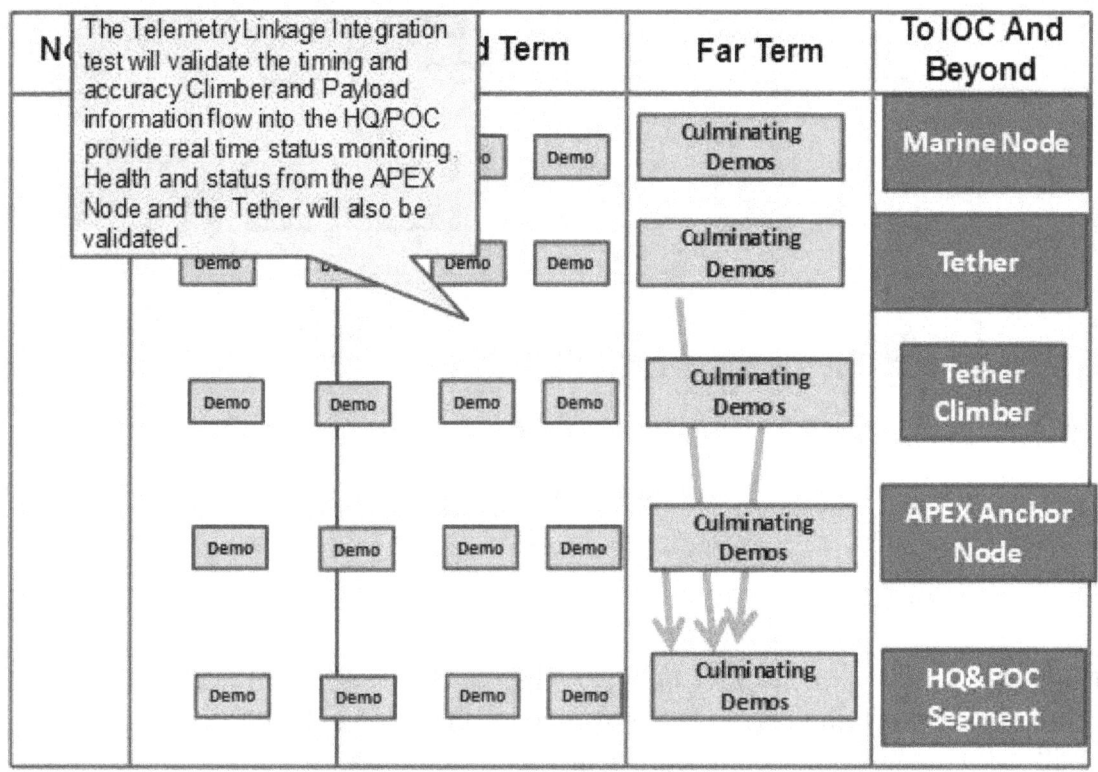

Figure 9-3 Telemetry Links

9.2.3 Payload Exchange.

The third integration objective is Payload exchange depicted in Figure 9-4 below. This is a physical handling objective and will largely be conducted on the ground. Many will see this as the validation part the payload acceptance function at the Marine Node, and it will be that. However, the transfer of the payload to the Climber and then from the Climber to the payload's delivery altitude are included in this integration task. The difficulty of this payload exchange is nested in the delivery altitude exchange. It is currently unclear how much video coverage and other real time monitoring will be needed at the delivery altitude. The Climber has extensive payload awareness responsibilities and this integration test will indeed "test" those responsibilities. Obviously, this integration testing will be preceded by a long series of ground pre-tests and some space based pathfinder tests.

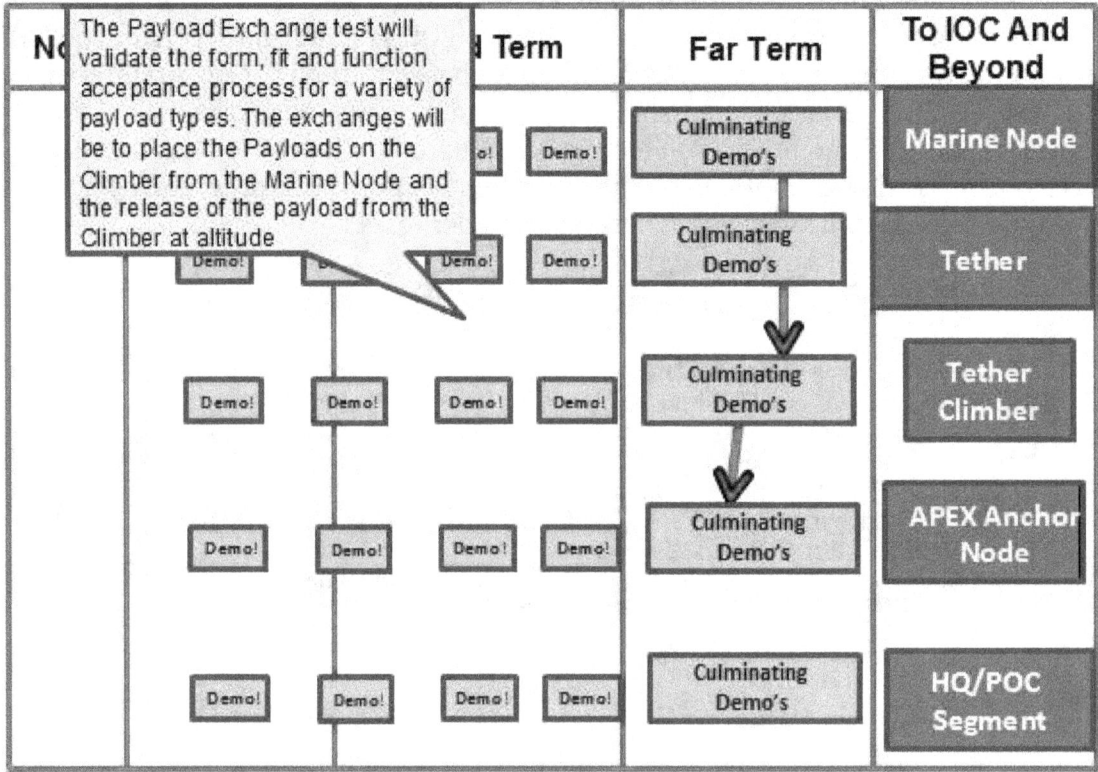

Figure 9-4   Payload Exchange

9.2.4   Operations Scenarios

The fourth integration objective is Operations Scenarios depicted in Figure 9-5 below. This integration testing is a thorough examination of how the overall Architecture conducts itself in the face of normal operations. It also is an examination of how well the operations team at HQ/ POC can deal with operational exigencies and anomalies. Part of these integration tests will be evaluation of the various "modes" of operation that will be formulated during the development process; the safe mode, power down mode, tether sever mode and so on. All such operational scenarios will be examined, and operating procedures will be re-documented to ensure safe and efficient Space Elevator operations. Here too, this integration testing will be preceded by a long series of ground pre-tests and space based pathfinder tests.

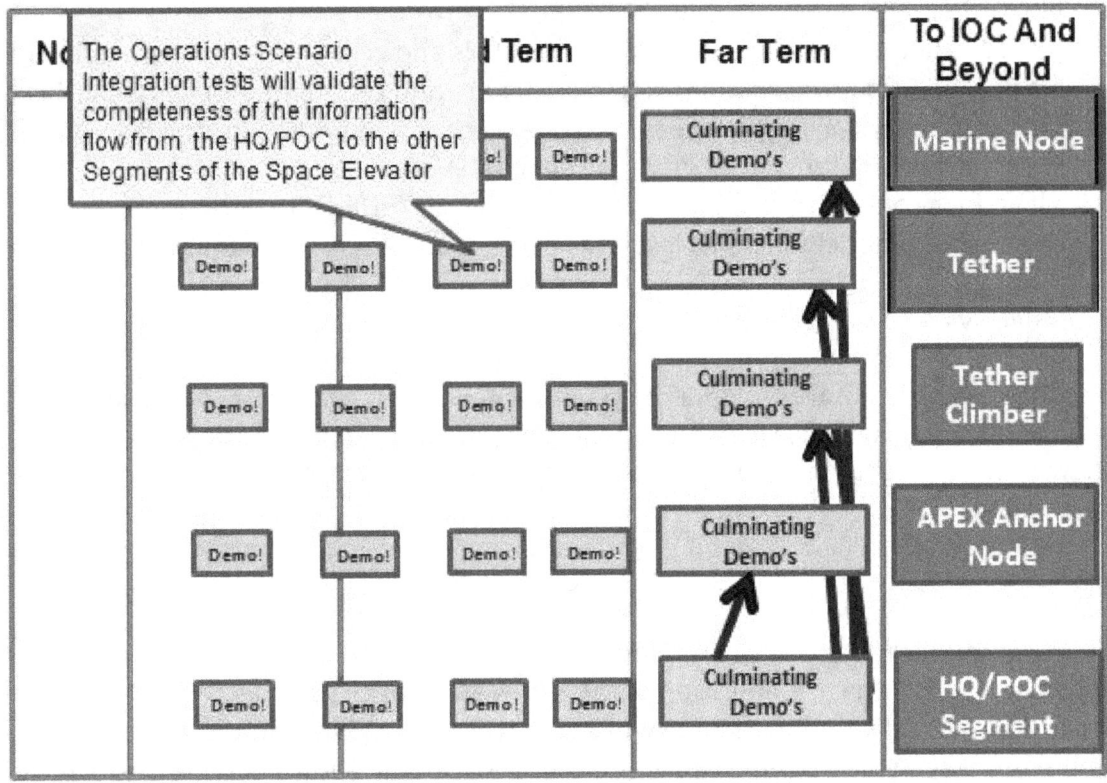

Figure 9-5 Operations Scenarios

9.3 Final Systems of Systems Integration Testing with Operations Checkout

After all the culminating demonstrations across the five segments, the validation for the customer could be a single full up system of systems integration test that shows the interplay across segments as well as within each segment. There is a concept surfacing within the space elevator community that sees a full-up test in space once the culminating demonstrations have been accomplished with individual segments. This would be designed to show to customers that the concept works in the environment of a space elevator, just at a smaller scale prior to commitment of the final investments. The concept is to have a long [estimate 1,000 kms] tether [of material less than required for the space elevator] being placed at 2,500 km altitude circular orbit. The initial spacecraft would resemble the future deployment satellite with both a smart satellite representing the Marine Node and a smart satellite representing the Apex Anchor, connected by a 1,000 km tether. The system of systems integration test would validate the following actions to ensure design maturity and concept viability:

- Placement in orbit (into LEO then transfered up to 2,500 km circular)
- Stabilization in circular orbit [2,500 km altitude]
- Deployment of tether under control [Apex Anchor deploys Marine Node under conditions similar to reality with HQPOC controlling]

- Full tether deployment [1,000 km length]
- Apex Anchor controls dynamics with thrusters and reel-in reel-out capabilities
- Marine Node controls dynamics with thrusters and reel-in reel-out capabilities
- Tether Climbers are deployed and climb up and down with mechanisms representing the full up design of the space elevator
- All actions controlled and monitored from the HQ/POC Primary center.
- Release of tether climbers as last tests [of course release to deorbit for debris cleanup]
- Move system to deorbit path.

Once the execution of this system of systems integration test is completed, all the serious questions of space elevator operations would have been addressed and answered showing a positive situation ready for commercial funding to deploy a space elevator.

9.4 Summary

This chapter has taken a high level look at the complex and busy conclusion of the design and preliminary integration of the space elevator. The space systems architects who look at this chapter will see that the flow of information seems confusing; but, in reality, illustrates the complexity of the situation. The engineers recognize that there are many preliminary actions necessary to ensure that future key achievements can be accomplished.

During this study, the group members have continually reminded themselves that we are to traverse the many various pathways multiple times. First we establish our vision of how the segments fit together. Then we establish pathways based upon technological maturity, segment knowledge, and testing methodologies. Each of the segment pathways is not independent and they each need information from the others as well as being necessary for the others to proceed. The development cycle is continuous and repetitive in many cases, but usually flowing to the right on a timeline.

10 Summary, Recommendations, and Vision

10.2 Summary

The bottom line summary for the development of a low cost space system transportation infrastructure is that it will be complex [with many varying degrees of difficulty]; but, the space elevator roadmap shows that it is definitely "a doable do." This chapter will lay out a basic summary of ideas across the document, discuss two recommendations for immediate implementation (as well as for longer term), and present a view of where we are going in the future. The process of road mapping enables us to understand our current place in time and technology readiness; and then, to develop a solid approach to taking the space elevator community to a recognizable future destination. The space system architecture destination, discussed in this ISEC document, is the IOC, depicted below in the Chase image:

Our Destination
The Initial Operational Capability (IOC) contains two space elevators, with separate Marine Nodes, Apex Anchors, 100,000 km usable tethers, and a single Headquarters and Primary Operations Center.

Figure 10-1 ISEC's Space Elevator IOC Architecture

This document has shown the IOC to be feasible. The culminating demonstrations are critical to its development. They each have multiple steps within each segment to achieve success. All paths are not isolated; but, are integrated with progress inside one segment dependent upon progress within another segment. The basic findings that have been reached within this study lead to the following conclusions:

- The Marine Node Segment will contain the most mature technologies inside the total transportation infrastructure. The one exception is that the tether terminus subsystem must be designed from TRLs in the range of 3 to 5.
- The Tether Segment is the critical technology for space elevators. Four very significant concerns are: 1) producability of long enough/strong enough material, 2) significant tensile strength material, 3) design for reparability, and 4) long enough life expectancy for commercially viable space elevators. Indeed, as the tether material goes – so goes the space elevator. The bottom line is that the tether, and especially the tether material, is the pacing item in the development of a space elevator.
- The Tether Climber Segment seems to be the most straight-forward system inside the space elevator architecture as most of the tasks have been accomplished within our space community over the last 50 years. The one exception to this is the gripping mechanisms of the tether climber to the tether.
- The Apex Anchor Segment is a very dynamic part of a space elevator. The ability to understand the natural motion, and then ensure stability, will be its most significant task. This would include the ability to understand the whole arena of tether dynamic stimulus such as: tether climbers, reeling-in/out of tether, forces from winds, movement of the Marine Node, and stability supplied by a large mass at the GEO Node. This requires the Apex Anchor to be an intelligent agent for the upper end of the tether.
- The HQ/POC Segment can be modeled as the brains of a space elevator system. As such, there will have to be an early, and substantive, version of the operations center to support the culminating demonstrations and systems integration tests. The demonstration pathway of the HQ/POC will indeed "touch" all the segments as well as ensure that all internal components are up to their tasking. Many of the early tests will validate the choices of automation or "human-in-the loop." Indeed, the final series of Grand Challenges will be conducted from an early implementation of the Primary Operations Center.

10.2 Recommendations

During the study looking at the IOC Architecture for the International Space Elevator Consortium, many factors surfaced and many issues were addressed. The study team believes the ISEC search for a roadmap to the IOC architecture is not finished. However, we do believe we have shown important steps in the process towards an IOC Architecture. As shown in the conclusions above, the ISEC has an opportunity to charge forward and continue its quest for both its vision and mission:

Our Vision is: A world with inexpensive, safe, routine, and efficient access to space for the benefit of all mankind.

Our Mission: The ISEC promotes the development, construction and operation of a Space Elevator (SE) Infrastructure as a revolutionary and efficient way to space for all humanity.

Within this context, the study team recommends the following three items:

1. Initiate the development, test and validation of a "high fidelity model" representing a space elevator system.
2. Endorse the current ISEC Research Plan [see at www.isec.org].
3. Engage in an active program to address the top five Grand Challenges discussed in this study report:
    a. High Fidelity Dynamics Modeling of the Tether as a system from the Marine Node to the Apex Anchor with GEO Node and tether climbers.
    b. Understand, measure, and improve the tensile strength of new materials with promise as future space elevator tethers.
    c. Study, design and test potential approaches for the gripping mechanism for tether climbers to move along the tether's 100,000 km length.
    d. Study robotic Apex Anchor functions with the anticipation that they will lead to a complex satellite. This should include the concern of operating and maintaining the Apex Anchor at a distance of 100,000 kms.
    e. Understand and design approaches for power to the tether climber for the hazardous first 40 km inside the atmosphere.

10.3 Future

After drafting the study report, and after a full year of refining the concepts and enhancing the presentations, the team sat down and tried to put the process and the future in perspective. As a team we all agreed to the following:

> ***We have definitely started down the path to a space elevator architecture! It is our intent that this ISEC trip be based upon a roadmap, which will take us towards the Initial Operational Capability (IOC) architecture, and will enable others to see our future.***

Appendix A   Architecture Attributes

Below are the key attributes of a Space Elevator.

Attribute	Edwards	IAA	Obayashi	ISEC
Length	100,000 km	100,000 km	96,000	100,000 km
Ribbon	1 meter wide, curved	1 meter wide, curved	Half Meter wide curved	1 meter wide, curved
#Cables	1 per FOP	1 per FOP	2 per FOP	1 per FOP
Material	CNT	CNT	CNT	CNT
Capacity	100 Pascal	25-35 MYuri	150 GPascal	25-35 MYuri
Climber		6MT	100 MT	6MT
Payload/s	20 MT	14 MT	79 MT	14 MT
Climber Power under 40 km	Laser	TBD Power Cord	Laser	TBD Power Cord
Climber over 40 km	Laser	Solar	Laser	Solar
Marine Node	Platform in E Pacific	Platform in E Pacific	Port extension from island	Platform in E Pacific

Table 1 Key Attributes

Appendix B Comparison of Current Architectures

Comparison of the published architectures for space elevators will be based upon a few important criteria: (1) Publishing and distribution of concepts that created significant steps in the development of space elevators, (2) the engineering level of detail was appropriate for the phasing of each report, and, (3) the presentation showed as much of the engineering as possible, at that time, enabling credibility for the development of space elevators. This led to a series of five architectures over the last 75 years. The first two were significant leaps in understanding:

- In 1960, Yuri Artsutanov presented a real approach visualizing how it could be achieved – a big leap from Tsiolkovsky's 1895 concept.
- Then, in 1974, Jerome Pearson resolved many issues with engineering calculations of tether strengths needed and approaches for deployment. This was once again a leap beyond Artsutanov's work and set the stage for the "modern design" for space elevators.
- Dr. Edwards established the current baseline for design of space elevator infrastructures at the turn of the century with his book entitled: "*Space Elevators*" [2002]. He established that the engineering could be accomplished in a reasonable time with reasonable resources. His baseline is solid; and, it was leveraged for the next two refinements of this transportation infrastructure concept.
- The International Academy of Astronautics leveraged Dr. Edwards' design and intervening ten years of excellent development work from around the globe. The 41 authors combined to improve the concept and establish new approaches, expanding the Edwards's baseline.
- The latest version of space elevator architectures is the recently released view by the Obayashi Corporation. Their set of assumptions of the study established stricter requirements and resulted in longer development with increased payload capacity.

The following sections will address each architecture as an increasingly improving infrastructure design. The differences will be emphasized as the reader moves forward from one architecture to the next. When the tether material develops, enabling the space elevator, the designers will take the best from each of the architectures and combine them together leading to a transportation infrastructure for low cost space access.

Origin & Architectures # I & II – Inventor's Concepts
Konstantin Tsiolkovsky, a Russian rocket scientist pioneered astronautics theory and conceptualized a building growing to GEO orbit in 1895.[8] He wrote about

---

[8] K. E. Tsiolkovski: Grezi o zemle I nebe, izd-vo ANSSSR, 35, 1959.

building a tower on the equator in his essay. The first player in the field to really deal with cables and layout a real concept was Yuri Artsutanov, who showed that you could stretch a cable from GEO downwards if the strength and lightness was significantly better than existed in 1960.[9] Figure B-1 shows the article.

In 1974, Jerome Pearson published his engineering calculations showing that the space elevator could be stable and built from a GEO orbit[10]. His image is in Figure A-2. As such, both Yuri Artsutanov and Jerome Pearson are considered co-inventors of the space elevator. These two architectures were remarkable in their time and set the stage for the next three versions.

An important change in the status quo came about in 1991 when Sumio Iijima discovered carbon nanotubes. He was with the NEC Fundamental Research laboratories.[11] These materials will "enable" the building of space elevators, when they are matured.

Figure B-1 To the Cosmos by Electric Train

---

[9] Y. Artsutanov : V Kosmos na Electrovoze, Komsomolskaya Pravda, July 31, 1960. Y. Artsutanov : To the Cosmos by Electric Train, SF magazine, Vol. 2, No. 2, pp121-123, 1961

[10] Jerome Pearson, The orbital tower: a spacecraft launcher using the Earth's rotational energy, Acta Astronautica. Vol. 2. pp. 785-799. Pergamon Press 1975. Printed in the U.S.A.

[11] S. Iijima : Helical microtubules of graphitic carbon, Nature, 354, pp56-58, 1991.

Figure B-2 Pearson and Artsutanov comparing notes

Figure B-3 Dr. Edwards' Basic Architecture

Architectures # III – Dr. Edwards' Architecture [2002]

The book "Space Elevators" by Dr. Brad Edwards and Eric Westling revolutionized the concept of a transportation infrastructure for space access with the realization that it could actually be constructed assuming that the carbon nanotube material developed as expected. The engineering design was solid and showed how the proposed space elevator interacted with the environment, including the hazards. The gravitational factors, radiation, heat/cold quandary, space debris, lightning and other atmospheric effects were all addressed and answered to an introductory level.

The basics of the design are shown in Figure B-3; and, a summary of the major items are shown below:

- Length: 100,000 km, anchored to an Earth terminus, as a large mass Marine Node, and connected to a large End Node [Could be an asteroid]
- Ribbon: Width-One meter, curved;
- Design: Woven with multiple strands and curved;
- Material: Carbon Nano-Tubes with 100 GPa strength at 1.3 gm/cm3
- Cargo: 20 metric ton payloads without humans
- Loading: Seven concurrent climbers on the ribbon
- Power Source: Terrestrial Lasers. See Figure B-4 below.
- Marine Node: Ocean going oil platform
- Operations Date: The space elevator can, and will, be produced in the near future [Ten years after start with mature materials].
- Construction Strategy: The first space elevator will be built from GEO; then, once the gravity well has been overcome, it will be replicated from the ground up.
- Price: $ 6 billion USD
- Cost per kg to GEO: $150 USD

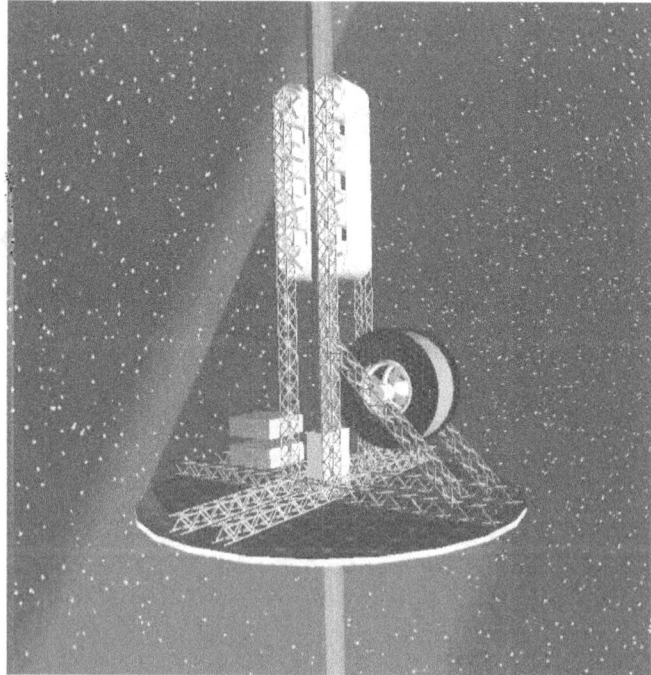

Figure B-4   Tether Climber with Laser Energy

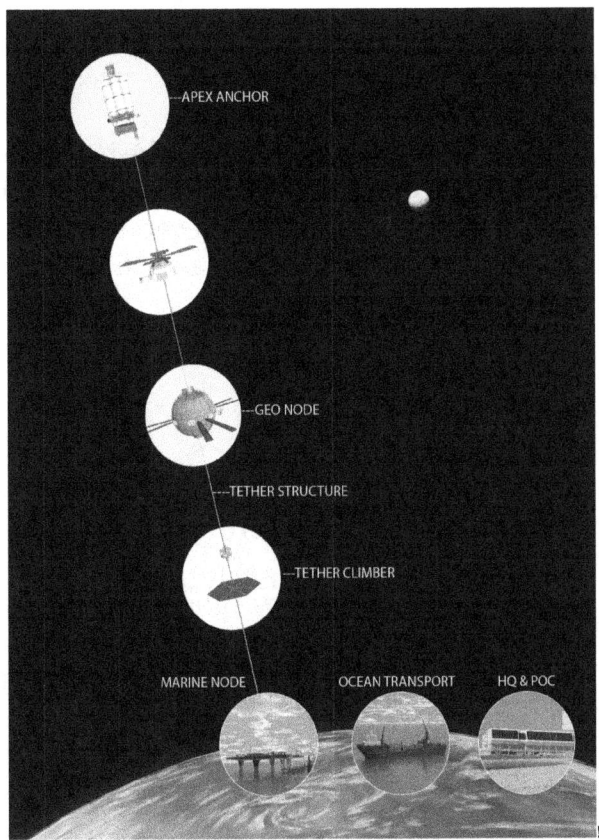

Figure B-5   IAA Architecture

Architectures # IV – International Academy of Astronautics [2013]

The IAA initiated a study in 2009 to evaluate the latest aspects of the potential low cost access to space infrastructure. The global effort included 41 authors, with 5 editors, from around the world and across many academic disciplines. The effort pulled together a 350 page book that addressed improvements to the potential architecture from Edwards' Baseline. [See Figure B-5 above] One consistent assumption was that human traffic on the space elevator would be at least a decade after the first few space elevators. The major differences are explained after a quick summary of the baseline configuration:

- Length: 100,000 km, anchored to floating Earth terminus, with a Marine Node connected to a large Apex Anchor.
- Ribbon: Width-One meter, curved;
- Design: Woven with multiple strands and curved;
- Material: Carbon Nano-Tubes with 25-35 MYuri at 1.3 gm/cm3
- Cargo: 14 metric ton payloads without humans [tether climber 6 MT]
- Loading: Seven concurrent payloads on the ribbon
- Power Source: Solar power after 1st 40 km
- Marine Node: Ocean going oil platform or retired aircraft carrier
- First 40 kms: box protection with power from an ultra-light cable.
- Alternative First 40 kms: High Stage One at 40 km altitude
- Apex Anchor: Based upon deployment satellite (with thrusters)

- Operations Date: The space elevator can, and will, be produced in the near future. [2035 operations start]
- Construction Strategy: The first space elevator will be built from GEO; then, once the gravity well has been overcome it will be replicated from the ground up.
- Architecture: Baseline is one replicating space elevator [used to produce all others] and then pairs sold to operating companies. Initial concept: three pairs operating around the world.
- Price: $ 13 billion for first pair, after replicator space elevator.
- Cost per kg to GEO: $ 500 USD

The improvements from older architectures were based in four significant parts:

Part I – Atmospheric Protection: The first forty kilometers of atmosphere are highly dangerous with all the normal factors such as lightning, high altitude winds, rain, sleet, and snow. As a preferred solution, the protective box climber was introduced. The initial tether climber would be no more than a storage box protecting the sensitive mission climber and solar arrays folded up inside. The image in Figure B-7 shows the deployment of folded solar arrays and tether climber as they climb out of the protective box. This enables the designers to protect mission climbers in a box in the atmosphere and as a free climber above the atmosphere. This simplifies the design and enables the mission climber to be much simpler in design. In addition, an alternate Marine Node design was considered which included the innovative approach to the same problem by developing a High Stage One [see Figure B-6 below]. This solution used mechanical energy to hold up a series of evacuated tubes leading to a work platform at 40 kms able to support over 400 metric tons without impacting the load on the space elevator tether.

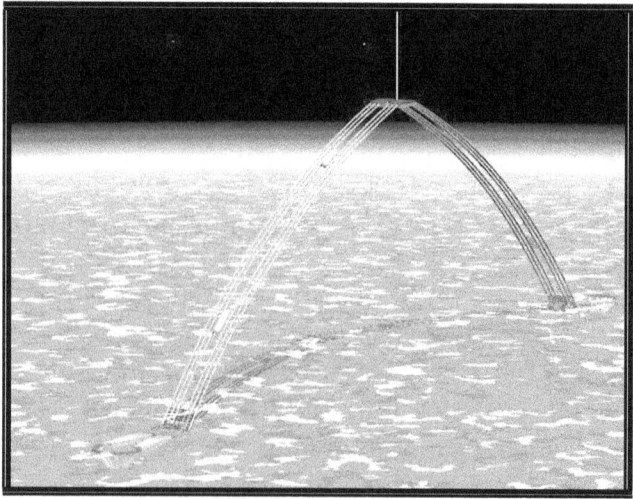

Figure B-6 High Stage One [Knapman image]

Part II – Lowering of needed tether strength level: Recent analyses of the whole problem showed that there were many trades to be evaluated. An example trade is the strength of materials to taper ratio with carrying capacity of each tether climber vs. material strength. A large study was conducted by Ben Shelef that ended up with a Space Elevator Feasibility Condition trading many factors against each other trying to optimize the design of the total system. As a result, the required level of material strength for a feasible space elevator turned out to be 30-35 MYuri. This is roughly 40 GPa [with density of material accounted for] with a safety factor of 40% [standard aerospace safety factor]. As a result, the needed material strength was about one third the numbers being looked for within the Edwards' approach. This moved the delivery date of sufficiently robust material forward. This enabled the program schedule to be drawn up as shown:

Figure B-7   Box Protection above Atmosphere[chasestudios.com]

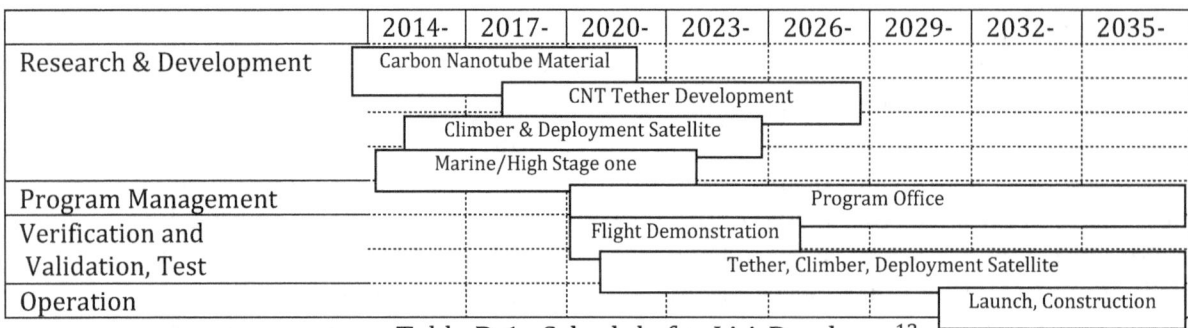

Table B-1   Schedule for IAA Roadmap[12]

Part III – Solar Power Only: The IAA study concluded that space elevator tether climbers could climb to GEO [and beyond] with solar power only. This assumption was developed after two studies were conducted. Again, Ben Shelef looked at the space elevator and the numbers resulted in the conclusion that solar power could work. The second analysis was conducted for the IAA and showed that if lightweight solar arrays for space that are projected by the current experts are, in fact, available on schedule [see documentation in IAA report], there would be sufficient power to raise a 20 Metric Ton climber, even in the heavy gravity at the 40 km altitude location. One key was that the tether climber now worked with the concept of constant power and varied its speed as the gravitational force lessened. Figure B- 8 below shows folded and deployed solar arrays [chasestudios.com]

---

[12] Minoru SATO, Akira TSUCHIDA, Review of the Space Elevator Research in Overseas, draft paper shared with author.

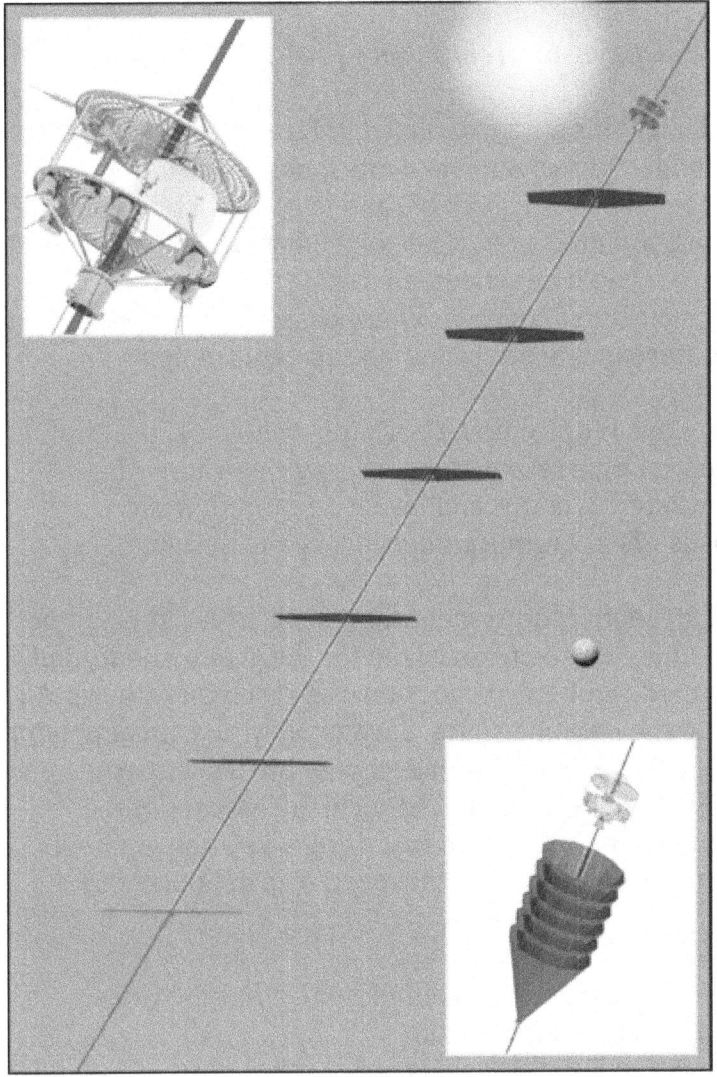

Figure B- 8   Folded and Deployed Solar Arrays [chasestudios.com]

The numbers are given for the needed area of solar arrays to provide the power required to move the tether climbers fast enough to make it to GEO within a week. Indeed, the arrays are large; but, the space industry has been dealing with asymmetric structures in space, for the last 60 years, and there shouldn't be any show stoppers when using solar arrays as the only power source above 40 kms.

Part IV – Apex Anchor is Smart: The previous architecture estimated that the counterweight could be an asteroid. The IAA study needs far more capability at the Apex Anchor; so, there is a demand for thrusters, computers, and communications at the upper tether terminus. This requires that the Apex Anchor be smart vs. dumb. The functional requirements include moderating the dynamic motion at the end point, reeling in and out to establish appropriate motion, and emergency force application if there is a tether sever. The Apex Anchor could contribute to the control of the remaining tether, if it was a smart node with thrusters.

Architecture V – Obayashi Architecture [2013][13]

The infrastructure that resulted from this study, depicted in Figures B-9, 10 and 11, is quite different and far more robust. However, when one looks at the initial requirements of their study, the results are reasonable and very consistent with the two previously conducted detailed architectures. The going in position for this most recent study was that human cargo would be scheduled during some of the first tether climber operations. This inclusion of humans was a basic requirement driving the design. The major differences are based upon this requirement.

Change One: Human transport – this requirement drove the need for 150 GPa of tether strength and the design of multiple tethers for each space elevator. In addition, the tether climbers consisted of many, and heavier, "cars" that carry people as well as heavier payloads. These requirements drove the design.

Resulting changes: The tether strength requirement was much larger, the number of cables per elevator increased, there is a tie to an island for the ground node, and there is a small aerodynamic shaped climber that goes through the atmosphere. All of these changes are additional to the baseline space elevator they plan on deploying up first. This baseline single tether is very similar in design to the basic space elevators of both Dr. Edwards and of the IAA study. As such, the essence of the developmental program is the same, with a more robust design building gradually, resulting from higher demands to support human transport. The details are as follows:

- Length: 96,000 km, anchored to an Earth terminus, with a Marine Node, connected to a large counterweight [12,500 ton]
- Ribbon: Width-half meter, curved; with 2 cables per carrier
- Design: With many cables leading to massive tether climbers
- Material: Carbon Nano-Tubes with 150 GPa capability
- Loading: Six concurrent payloads on the ribbon [both up/down]
- Power Source: Laser power from ground or space
- Marine Node: Port extension from island, 40 million MT, 400 m diameter
- Climber: 100 MT, with 79 MT payloads
- GEO station: 66 modules at 4,000 MT
- Operations Date: The space elevator can, and will, be produced in the near future. [2055 operations]
- Construction Strategy: The first space elevator will be built from GEO: then, once the gravity well has been overcome, it will be replicated from the ground up. First cable in 17 years, large capability after 18 years of building up cable.
- Architecture: One large space elevator with maximum capability
- Price: $ 100 billion USD
- Cost per kg to GEO: $ 50-100 USD

---

[13] Permission to use images was given by Obayashi Corporation Oct 2014.

Fig.B-9 : View of the human space elevator (Illustration)

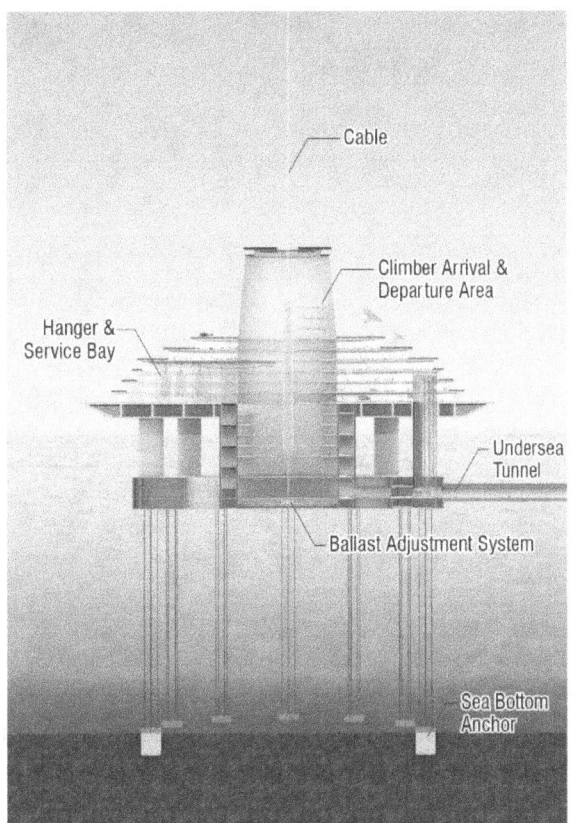

Fig. B-10 Earth Port (main facility)

81

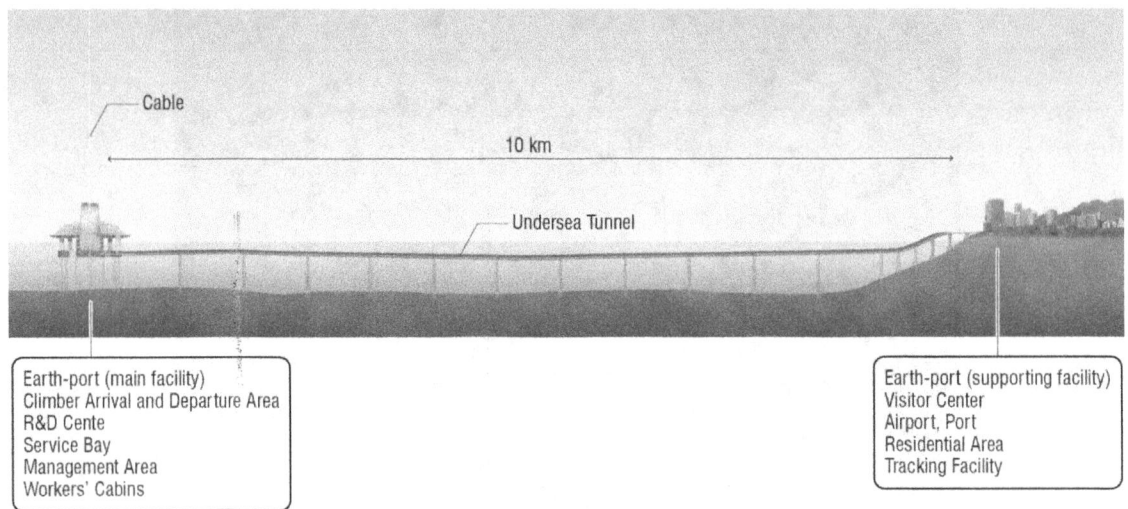

Fig B-. 11 Earth Port (onshore main facility and offshore supporting facility)

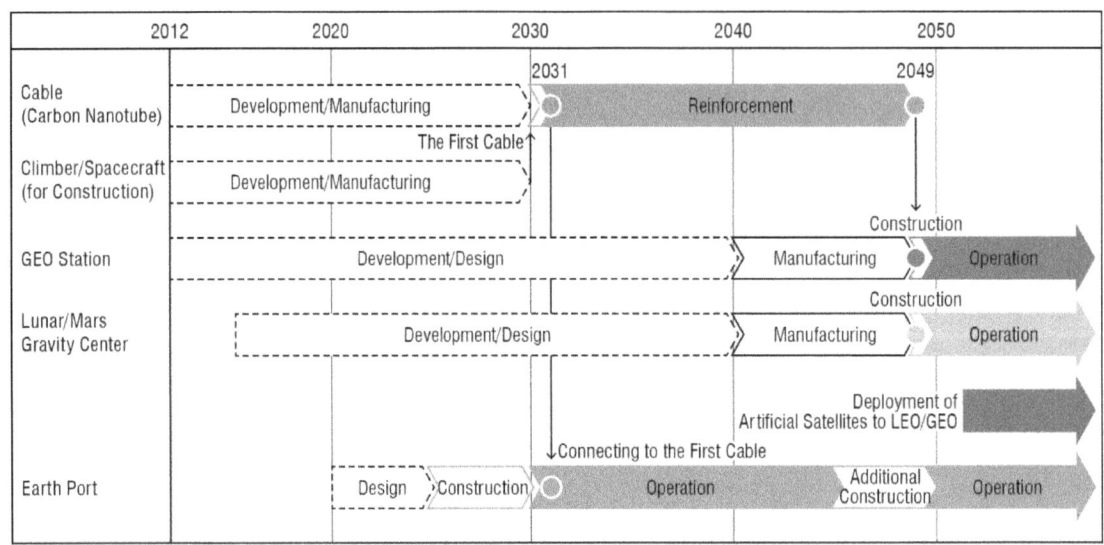

Fig. B-12 Construction schedule

References:
K. E. Tsiolkovski: Grezi o zemle I nebe, izd-vo ANSSSR, 35, 1959.

Y. Artsutanov : V Kosmos na Electrovoze, Komsomolskaya Pravda, July 31, 1960. Y. Artsutanov : To the Cosmos by Electric Train, SF magazine, Vol. 2, No. 2, pp121-123, 1961

Jerome Pearson, The orbital tower: a spacecraft launcher using the Earth's rotational energy, Acta Astronautica. Vol. 2. pp. 785-799. Pergamon Press 1975. Printed in the U.S.A.

Edwards, B. & E. Westling, The Space Elevator, BC Edwards, Houston, 2002.

Swan, P, D. Raitt, C. Swan, R. Penny, J. Knapman, "Space Elevators: An Assessment of

the Technological Feasibility and the Way Forward, Virginia Edition Publishers, Houston, 2013.
Ishikawa, Yoji, The Space Elevator Construction Concept, Obayashi Corporation, 2013, IAC-13-D4.3.6.

Appendix C   International Space Elevator Consortium

Who We Are
The International Space Elevator Consortium (ISEC) is composed of individuals and organizations from around the world who share a vision of humanity in space.

Our Vision
A world with inexpensive, safe, routine, and efficient access to space for the benefit of all mankind.

Our Mission
The ISEC promotes the development, construction and operation of a space elevator infrastructure as a revolutionary and efficient way to space for all humanity.

What We Do
- Provide technical leadership promoting development, construction, and operation of space elevator infrastructures.
- Become the "go to" organization for all things space elevator.
- Energize and stimulate the public and the space community to support a space elevator for low cost access to space.
- Stimulate science, technology, engineering, and mathematics (STEM) educational activities while supporting educational gatherings, meetings, workshops, classes, and other similar events to carry out this mission.

A Brief History of ISEC
The idea for an organization like ISEC had been discussed for years; but, it wasn't until the Space Elevator Conference in Redmond, Washington, in July of 2008, that things became serious. Interest and enthusiasm for a space elevator had reached an all-time peak; and, with Space Elevator conferences upcoming in both Europe and Japan, it was felt that this was the time to formalize an international organization. An initial set of directors and officers were elected; and, they immediately began the difficult task of unifying the disparate efforts of space elevator supporters worldwide.

ISEC's first Strategic Plan was adopted in January of 2010 and it is now the driving force behind ISEC's efforts. This Strategic Plan calls for adopting a yearly theme to focus ISEC activities. (For 2010, the theme was "Space Elevator Survivability -- Space Debris Mitigation.") In 2010, ISEC also announced the first annual Artsutanov and Pearson prizes to be awarded for "exceptional papers that advance our understanding of the Space Elevator." Because of our common goals and hopes for the future of mankind off--planet, ISEC became an Affiliate of the National Space Society in August of 2013.

Our Approach

ISEC's activities are pushing the concept of space elevators forward. These cross all the disciplines and encourage people from around the world to participate. The following activities are being accomplished in parallel:

- CLIMB – This annual peer reviewed journal invites and evaluates papers and presents them in an annual publication with the purpose of explaining technical advances to the public. The first issue of CLIMB was dedicated to Mr. Yuri Artsutanov (a co-inventor of the space elevator concept); and, the second issue was dedicated to Mr. Jerome Pearson (another co--inventor). CLIMB is scheduled for publication in July.
- Yearly conference – International space elevator conferences were initiated by Dr. Brad Edwards in the Seattle area in 2002. Follow--on conferences were in Santa Fe (2003), Washington DC (2004), Albuquerque (2005/6 –smaller sessions), and Seattle (2008 to the present). Each of these conferences had multiple discussions across the whole arena of space elevators with remarkable concepts and presentations. Recent conferences have been sponsored by Microsoft, the Seattle Museum of Flight, the Space Elevator Blog, the Leeward Space Foundation, and ISEC.
- Yearlong technical studies – ISEC sponsors research into a focused topic each year to ensure progress in a discipline within the space elevator project. The first such study was conducted in 2010 to evaluate the threat of space debris. The second study, and resulting report, focused on space elevator operations. The 2013 study focused upon tether climber designs. The 2014 topic is Space Elevator Architectures and Roadmaps. There are two topics chosen for 2015; Marine Node Design Considerations and Status of Tensile Strength materials development. The products from these studies are reports that are published to document progress in the development of space elevators.
- International cooperation – ISEC supports many activities around the globe to ensure that space elevators keep progressing towards a developmental program. International activities include coordinating with the two other major societies focusing on space elevators: the Japanese Space Elevator Association and EuroSpaceward. In addition, ISEC supports symposia and presentations at the International Academy of Astronautics and the International Astronautical Federation Congress each year.
- Competitions – ISEC has a history of actively supporting competitions that push technologies in the area of space elevators. The initial activities were centered on NASA's Centennial Challenges called "Elevator: 2010." Inside this were two specific challenges: Tether Challenge and Beam Power Challenge. The highlight came when Laser Motive won $900,000 in 2009, as they reached one kilometer in altitude racing other teams up a tether suspended from a helicopter. There were also multiple competitions where different strengths of materials were tested going for a NASA prize – with no winners. In addition, ISEC supports the educational efforts of various organizations, such as the LEGO space elevator

climb competition at our Seattle conference. Competitions have also been conducted in both Japan and Europe.
- Publications – ISEC publishes a monthly e--Newsletter, its yearly study reports and an annual technical journal [CLIMB] to help spread information about space elevators.
- Reference material – ISEC is building a Space Elevator Library, including a reference database of Space Elevator related papers and publications.
- Outreach – People need to be made aware of the idea of a space elevator. Our outreach activity is responsible for providing the blueprint to reach societal, governmental, educational, and media institutions and expose them to the benefits of space elevators. ISEC members are readily available to speak at conferences and other public events in support of the space elevator. In addition to our monthly e--Newsletter, we are also on Facebook, Linked In, and Twitter.
- Legal – The space elevator is going to break new legal ground. Existing space treaties may need to be amended. New treaties may be needed. International cooperation must be sought. Insurability will be a requirement. Legal activities encompass the legal environment of a space elevator -- international maritime, air, and space law. Also, there will be interest within intellectual property, liability, and commerce law. Starting work on the legal foundation well in advance will result in a more rational product.
- History Committee – ISEC supports a small group of volunteers to document the history of space elevators. The committee's purpose is to provide insight into the progress being achieved currently and over the last century.
  - Research Committee – ISEC is gathering the insight of researchers from around the world with respect to the future of space elevators. As scientific papers, reports and books are published, the research committee is pulling together this relative progress to assist academia and industry to progress towards an operational space elevator infrastructure. For more visit http://isec.org/index.php/about-isec/isec-research-committee

ISEC is a traditional not-for-profit 501 (c) (3) organization with a board of directors and four officers: President, Vice President, Treasurer, and Secretary. In addition, ISEC is closely associated with the conference preparation team and other volunteer members.

Address: ISEC, 709--A N. Shoreline Blvd, Mountain View, CA 94043 (630) 240--4797 / inbox@isec.org / www.isec.org

Appendix D TAI's Architecture and Roadmap Approach

The recurring theme of Technology, Architectures and Integration, LLC., (TAI) experience within developmental programs was that solid and thorough testing produced the technical information which resolved design issues and performance challenges. This realization came from Mr. Fitzgerald's years of participation within several USAF space and missile development programs. He has worked with many operational space programs, as well as missile warning, and Intercontinental Ballistic Missile systems. His specialty is gaining an early understanding of the great potential gained from new systems and their acquisition processes.

The heart of his company's approach is to bring solid empirical information to resolve design and performance issues. TAI assists in the application of technology to the needed customer solutions to develop a new system or upgrade. With the rate of technology turnover in today's technical marketplace, solid technology insertion and application approaches are necessary for long-term business and performance success. TAI has access to many resources that can be applied to new systems for a variety of customers. TAI will help find achievable architectures and roadmaps for your exciting endeavor. Our approach is depicted below.

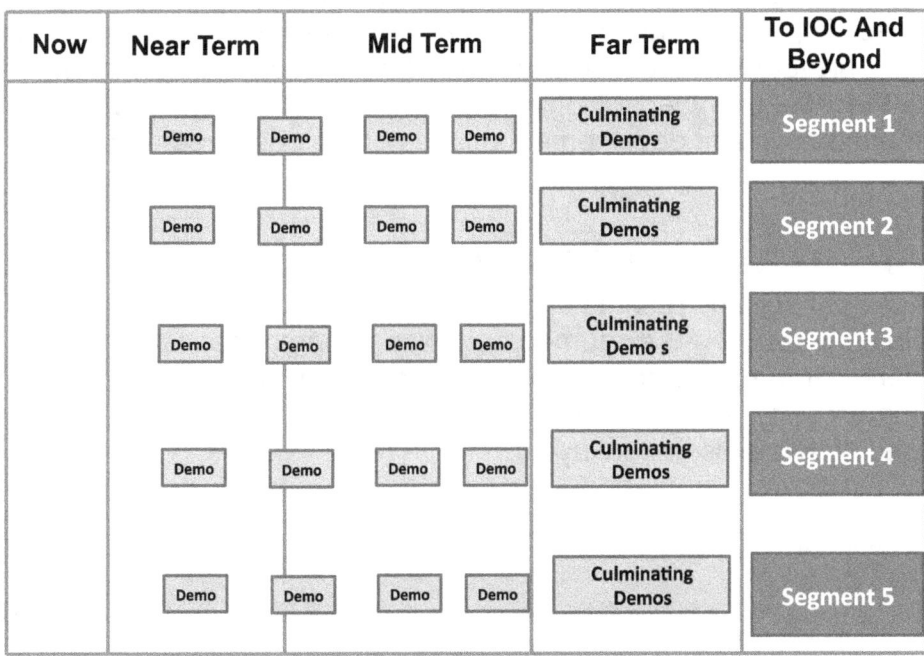

Figure D-1   TAI Approach

Objectives

On the last day of the 2014 Space Elevator conference in Seattle, the ISEC Roadmaps and Architecture study leads held a workshop with the attendees of the conference. The workshop had two purposes.

Purposes:

- To define the ISEC Space Elevator Architecture in five discrete segments: Climber, Tether, Marine Node, HQ/POC and Tether Tenants. Of these, the first three, representing tether operations, were discussed in detail.
- To seek feedback from the attendees regarding demonstrations and success criteria for functions within the reviewed segments

Segments discussed and reviewed

The workshop briefer presented graphics for the three four segments within the space elevator architecture. The graphics (from Chapters 4-7 of this report) portrayed the path along which segments must move on their way to preliminary and then detailed implementation plans. For the development engineers, these implementation plans are the needed series of design efforts to build the space elevator. To move toward these plans, each segment must demonstrate that necessary technologies and engineering solutions will be available for the design and development phases of the portrayed roadmap.

Observations and conclusions

The idea of exposing months of hard work to the public for inspection is always a little embarrassing. As it turned out I was not embarrassed; the attendees were warm and welcoming in their feedback. It was a wonderful, humbling moment and I am proud to have been a part of it. The engineering of the ISEC Space Elevator is underway!